Wissenschaftliche Reihe Fahrzeugtechnik Universität Stuttgart

Reihe herausgegeben von

Michael Bargende, Stuttgart, Deutschland

Hans-Christian Reuss, Stuttgart, Deutschland

Jochen Wiedemann, Stuttgart, Deutschland

Das Institut für Fahrzeugtechnik Stuttgart (IFS) an der Universität Stuttgart erforscht, entwickelt, appliziert und erprobt, in enger Zusammenarbeit mit der Industrie, Elemente bzw. Technologien aus dem Bereich moderner Fahrzeugkonzepte. Das Institut gliedert sich in die drei Bereiche Kraftfahrwesen, Fahrzeugantriebe und Kraftfahrzeug-Mechatronik. Aufgabe dieser Bereiche ist die Ausarbeitung des Themengebietes im Prüfstandsbetrieb, in Theorie und Simulation. Schwerpunkte des Kraftfahrwesens sind hierbei die Aerodynamik, Akustik (NVH), Fahrdynamik und Fahrermodellierung, Leichtbau, Sicherheit, Kraftübertragung sowie Energie und Thermomanagement – auch in Verbindung mit hybriden und batterieelektrischen Fahrzeugkonzepten. Der Bereich Fahrzeugantriebe widmet sich den Themen Brennverfahrensentwicklung einschließlich Regelungs- und Steuerungskonzeptionen bei zugleich minimierten Emissionen, komplexe Abgasnachbehandlung, Aufladesysteme und -strategien, Hybridsysteme und Betriebsstrategien sowie mechanisch-akustischen Fragestellungen. Themen der Kraftfahrzeug-Mechatronik sind die Antriebsstrangregelung/Hybride, Elektromobilität Bordnetz und Energiemanagement, Funktions- und Softwareentwicklung sowie Test und Diagnose. Die Erfüllung dieser Aufgaben wird prüfstandsseitig neben vielem anderen unterstützt durch 19 Motorenprüfstände, zwei Rollenprüfstände, einen 1:1-Fahrsimulator, einen Antriebsstrangprüfstand, einen Thermowindkanal sowie einen 1:1-Aeroakustikwindkanal. Die wissenschaftliche Reihe „Fahrzeugtechnik Universität Stuttgart" präsentiert über die am Institut entstandenen Promotionen die hervorragenden Arbeitsergebnisse der Forschungstätigkeiten am IFS.

Reihe herausgegeben von

Prof. Dr.-Ing. Michael Bargende
Lehrstuhl Fahrzeugantriebe
Institut für Fahrzeugtechnik Stuttgart
Universität Stuttgart
Stuttgart, Deutschland

Prof. Dr.-Ing. Hans-Christian Reuss
Lehrstuhl Kraftfahrzeugmechatronik
Institut für Fahrzeugtechnik Stuttgart
Universität Stuttgart
Stuttgart, Deutschland

Prof. Dr.-Ing. Jochen Wiedemann
Lehrstuhl Kraftfahrwesen
Institut für Fahrzeugtechnik Stuttgart
Universität Stuttgart
Stuttgart, Deutschland

Weitere Bände in der Reihe http://www.springer.com/series/13535

Sebastian Hann

A Quasi-Dimensional SI Burn Rate Model for Carbon-Neutral Fuels

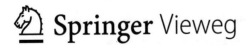 Springer Vieweg

Sebastian Hann
Institute of Automotive Engineering
(IFS), Chair in Automotive Powertrains
University of Stuttgart
Stuttgart, Germany

Zugl.: Dissertation Universität Stuttgart, 2020

D93

ISSN 2567-0042   ISSN 2567-0352   (electronic)
Wissenschaftliche Reihe Fahrzeugtechnik Universität Stuttgart
ISBN 978-3-658-33231-0   ISBN 978-3-658-33232-7   (eBook)
https://doi.org/10.1007/978-3-658-33232-7

This Springer Vieweg imprint is published by the registered company Springer Fachmedien Wiesbaden GmbH part of Springer Nature.
The registered company address is: Abraham-Lincoln-Str. 46, 65189 Wiesbaden, Germany

Preface

My deep gratitude goes to Prof. Dr.-Ing. M. Bargende for the opportunity to realize this thesis during my tenure as a research associate at the Institute of Automotive Engineering (IFS) at the University of Stuttgart. I truly appreciate his outstanding support, guidance and the numerous inspiring discussions.

I want to thank Prof. Dr. sc. techn. T. Koch for his interest in my work and for joining the doctoral committee.

I am extremely grateful to all my colleagues at the IFS and FKFS for creating an unprecedented working environment. In particular, I like to thank Dr. M. Grill for his valuable guidance and endless support. Furthermore, special thanks goes to L. Urban for his friendship and initially sparking my interest in the subject, S. Fritsch for providing me with 3D CFD data and sharing his insights in turbulence effects, Dr. D. Schmidt for sharing his knowledge of reaction kinetics, Dr. D. Rether for his patient introduction into Fortran coding and H. Fußhoeller for his support in organizational matters. I also like to acknowledge the student works of M. Schmid, R. Nußbaumer and S. Crönert, who I was very happy to supervise.

My appreciation also extends to the working group, led by Dr. M. Schenk and Dr. F. Altenschmidt, as well as the companies that supported the research task within the project "Methane Fuels II" defined and financed by the Research Association for Combustion Engines (FVV) e.V.. Deep gratitude goes to my project partner M. Eberbach, who kindly provided me with measurement data and let me join him at the engine test bench.

Furthermore, I want to thank my parents for their continuous support, helping me through stressful times and allowing me to focus on the essential.

Stuttgart   Sebastian Hann

Contents

Figures

Tables

Abbreviations

0D	Zero-dimensional
1D	One-dimensional
3D	Three-dimensional
ASTM	American Society for Testing and Materials
AWC	Averaged working cycle
CAD	Crank angle degree, computer aided design
CCV	Cycle-to-cycle variations
CFD	Computational fluid dynamics
CH_4	Methane
CNG	Compressed natural gas
CO	Carbon monoxide
CO_2	Carbon dioxide
COV	Coefficient of variation
CR	Compression ratio
CRECK	Chemical Reaction Engineering and Chemical Kinetics Lab, Politecnico di Milano
DI	Direct injection
DIN	Deutsches Institut für Normung (German Institute for Standards
DL	*Darrieus-Landau*
DMC+	65 vol-% dimethyl carbonate + 35 vol-% methyl formate
E85	Gasoline with 85 vol-% ethanol
EGR	Exhaust gas recirculation
EN	Europäisch Norm (European Standard)
FKFS	Research Institute of Automotive Engineering and Vehicle Engines Stuttgart

FVV	Research Association for Combustion Engines e. V.
GT	Gamma Technologies, LLC.
H	Hydrogen atom
H/C	Hydrogen-to-carbon
HO_2	Hydroperoxyl radical
HRR	Heat release rate
IFS	Institute of Automotive Engineering, University of Stuttgart
IMAT	Intake manifold air temperature
IMEP	Indicated mean effective pressure
ISFC	Indicated specific fuel consumption
KO	Knock onset
LLNL	Lawrence Livermore National Laboratory
LNG	Liquified natural gas
MeFo	Methyl formate
MFBxx	xx % Mass fraction burnt
MON	Motor octane number
NUI	National University of Ireland
O_2	Oxygen
OH	Hydroxyl radical
PDF	Probability density function
PFI	Port fuel injection
PRF	Primary reference fuel
PTA	Pressure trace analysis
QD	Quasi-dimensional
rel.	Relative

RON	Research octane number
SI	Spark ignition
Sim.	Simulation
SWC	Single working cycle
TDCF	Top dead center firing
TKE	Turbulent kinetic energy
TRF	Toluene reference fuel
Vari.	Variation

Symbols

Latin Letters

A	Surface area	m^2
a	Calibration parameter, local u'	-
A	Calibration parameter, τ	ms
$a_{\delta_{\mathrm{L}}}$	Calibration parameter, δ_{L}	-
$a_{l_{\mathrm{int}}}$	Calibration parameter, l_{int}	-
A_{piston}	Piston surface area	m^2
a_{τ}	Calibration parameter, τ	-
a_{u}	Calibration parameter, u_{E}	-
B_{i}	Calibration parameter, s_{L}	bar
b_{yH2O}	Calibration parameter, s_{L}	-
c	Calibration parameter, s_{L}	-
c_{p}	Specific heat capacity, constant pressure	J/kg/K
c_{v}	Specific heat capacity, constant volume	J/kg/K
c_{yH2O}	Calibration parameter, s_{L}	K
dQ_{b}	Heat release rate	J/CAD
d_{yH2O}	Calibration parameter, s_{L}	-
E_{A}	Activation energy	J/mol
E_{i}	Calibration parameter, s_{L}	K
F	Calibration parameter, s_{L}	m/s
G	Calibration parameter, s_{L}	K
$H_{\mathrm{u,mix,grav.}}$	Gravimetric lower heating value of the mixture	J/kg
I_{k}	*Livengood-Wu* Integral	-
Ka	*Karlovitz* number	-
Le	*Lewis* number	-
l_{int}	Integral length scale	m
l_{K}	*Kolmogorov* length scale	m
L_{min}	Stoichiometric air-fuel ratio	-
l_{T}	*Taylor* microscale	m
m	Calibration parameter, local u' and s_{L}	-
Ma	*Markstein* number	-

$max(u'_{fac})$, *target*	Calibration parameter, local u'	-
m_b	Burnt mass	kg
m_E	Entrained mass	kg
m_F	Mass in the flame front	kg
n	Calibration parameter, s_L	-
n_a	Calibration parameter, s_L	-
n_{EGR}	Calibration parameter, s_L	-
p	Pressure	bar
r	Calibration parameter	-
R	Flame radius, gas constant	m or J/mol/K
r_{bore}	Radius of cylinder bore	m
r_{flame}	Flame radius	m
r_{max}	Maximum flame radius	m
r_{rel}	Relative flame radius	-
s_f	Stretched flame speed	m/s
s_L	Laminar flame speed	m/s
$s_{L,0}$	Φ-dependent s_L in *Gülder* / *Heywood*	m/s
$s_{L,H2O}$	Laminar flame speed incl. influence of water injection	m/s
sp_{hor}	Horizontal spark plug position, calibration parameter	m
s_T	Turbulent flame speed, see u_E	m/s
S_x	Calibration parameter, s_L	-
t	Time	s
t_0	Start time, lower limit of *Livengood-Wo* integral	s
T_0	Reaction zone temperature	K
T_{ad}	Adiabatic flame temperature	K
T_b	Temperature (burnt gas)	K
$TDC_{clearance}$	Top dead center clearance, calibration parameter	m
t_E	End time, upper limit of *Livengood-Wo* integral	s
T_u	Temperature (unburnt gas)	K
u^0	Normal local flame speed	m/s

u_E	Entrainment velocity, see s_T	m/s
u'	Turbulent fluctuation velocity	m/s
u'_{eff}	Effective u'	m/s
u'_{fac}	Correction factor to consider local u'	-
u_t	Tangential local flame speed	m/s
u_u^0	Normal local flame speed, unburnt	m/s
V_b	Burnt volume	m^3
$V_{b,rel}$	Relative burnt volume	-
V_{cyl}	Cylinder volume	m^3
V_d	Displacement volume per cylinder	m^3
x	Distance	m
X	Mole fraction	-
Y_{EGR}	Stoichiometric EGR mass fraction	-
Y_{H2O}	Mass fraction of injected water	-
Y_{react}	Reactive mass fraction	-
Z^*	Mixture fraction	-
Z^*_{st}	Stoichiometric mixture fraction	-

Greek Letters

α	Thermal diffusivity and calibration parameter for τ	m^2/s or -
β	Calibration parameter, τ	-
χ_{Taylor}	*Taylor*-parameter	-
χ_{ZS}	Calibration parameter, CCV model	-
δ_L	Laminar flame thickness	m
$\delta_{L,dT,max}$	Laminar flame thickness at maximum temperature gradient	m
δ_{L,T_0}	Laminar flame thickness, evaluated at inner layer	m
δ_M	*Markstein* length	m
η	Dynamic viscosity	kg/m/s
η_i	Indicated engine efficiency	-
γ	Calibration parameter, τ	-
κ	Stretch rate	1/s

λ	Air-fuel equivalence ratio, thermal conductivity	- and W/m/K
ν	Kinematic viscosity	m^2/s
ν_L	Laminar flame velocity	m/s
ν_T	Turbulent kinetic viscosity	m^2/s
Φ	Air-fuel equivalence ratio	-
φ_{ZS}	Calibration parameter, CCV model	-
ρ_b	Density (burnt gas)	kg/m^3
ρ_u	Density (unburnt gas)	kg/m^3
τ	Ignition delay time, characteristic time scale of a single step reaction	s
τ_F	Chemical time scale	s
τ_K	*Kolmogorov* time scale	s
τ_L	Characteristic burn-up time	s

Indices

0	Base, reference, normal
ad	Adiabatic
b	Burnt
cyl	Cylinder
E	Entrainment
eff	Effective
EGR	Exhaust gas recirculation
Eth	Ethane
F	Flame front
f	Stretched flame
fac	Factor
fl	Flame
grav.	Gravimetric
hor	Horizontal
i	Inside
int	Integral
j	Mean location in turbulent flame front
K	*Kolmogorov*
L	Laminar

M	*Markstein*
max	Maximum
meas	Measured, measurement
mix	Mixture
o	Outside
r	Root
react	Reactive
ref	Reference
rel	Relative
st	Stoichiometric
T	Turbulent, *Taylor*
t	Tangential, tip
u	Unburnt

Abstract

As it is essential that action is taken to counteract climate change, all possible options must be evaluated for reducing CO_2 emissions in each sector, while taking a product's entire life cycle into account. Furthermore, specific solutions for each individual (sub-) sector could be advantageous compared to only a few but all-embracing solutions. For the transport sector, this translates to the unbiased evaluation of different concepts, such as battery electric vehicles, fuel cells, hybrid powertrains or optimized combustion engines, powered by alternative or synthetic fuels. For the evaluation of the latter, 0D/1D engine simulation as a complementary method to engine test bed investigations offers a fast and cost-efficient approach. For this, reliable models are needed to predict the influence of both the fuel and a variation of boundary conditions – such as air-fuel ratio, high rates of residual gas or water injection – on engine combustion. The development and validation of such a model on the basis of existing models was the aim of this thesis.

To reach this target, fundamental research results concerning the fuel influence on turbulent flame propagation were taken from the literature. These showed a general fuel influence on flame wrinkling as well as its dependency on the air-fuel ratio. However, the investigations of fundamental research are limited to boundary conditions with little relevance for engine operation. Therefore, their application to engine operation is questionable. In consequence, test bed measurements of an engine powered by different fuels and operated at varying air-fuel ratios were investigated, but no fuel influence on flame wrinkling was observed. With that, a prediction of fuel influence in engine simulation should be possible by introducing a fuel-dependent model for the laminar flame speed alone.

Since measurements of the laminar flame speed are also limited to boundary conditions irrelevant for engine operation, reaction kinetics calculations were used as a basis for the model development. As fuels, methane, CNG, methanol, ethanol, gasoline, hydrogen, methyl formate and DMC+ (65 vol-% dimethyl carbonate, 35 vol-% methyl formate) were considered. The reaction kinetics calculation results were approximated using an improved, semi-phenomenological model. On this basis, a model for the laminar flame thickness was also developed, as this is needed for the calculation of the turbulent flame speed. The model for this calculation was also evaluated and replaced by a more sophisticated one. Subsequently, the calculation of the characteristic burn-up time, a key feature of the entrainment approach, was reworked as well as interpreted phenomenologically and mathematically. The latter illustrated the influence of single working cycle heat release rates on that of an averaged working cycle. This not only highlights the linking of simulated heat release rates to a certain level of cycle-to-cycle variations, but also questions the necessity for an additional model covering the early phase of combustion. Further improvement of the burn rate model was achieved by introducing a mathematical model to account for the local distribution of the turbulent kinetic energy. This model was developed on the basis of 3D CFD calculations. The improved burn rate model was then linked to an existing cycle-to-cycle variations model.

Six different engines were used to validate the burn rate model. The engines were operated using different fuels at varying air-fuel ratios as well as different rates of EGR and water injection. The validation results justified the approach to cover the fuel influence on combustion by a change in laminar flame speed and thickness alone. With that, the improved burn rate model not only achieves the intended predictive abilities without the need for a model recalibration, but also allows for the development of future engine concepts.

The approach to cover the fuel influence in engine simulation by accounting for chemical effects, which corresponds to the laminar flame speed and thickness when related to combustion, was also evaluated briefly for engine knock by expanding a model of the ignition delay time to cover additional fuels. For both single and averaged working cycles, the approach allowed a reliable prediction of the fuel influence on knock onset and the knock limit. Furthermore, the investigation allowed the influence of top land gas discharge on knock occurrence to be discarded. This supports the theory of knock occurring in large end gas pockets.

Kurzfassung

Das Ergebnis dieser Arbeit ist ein quasi-dimensionales Modell, welches die Vorhersage der Brennverlaufsänderung bei einer Variation des Kraftstoffs von Benzin über CNG bis hin zu alternativen und synthetischen Kraftstoffen wie Methanol, Ethanol, Wasserstoff, Methylformiat oder DMC+ in der 0D/1D-Motorsimulation ermöglicht, ohne dabei neu abgestimmt werden zu müssen. Diese Vorhersagefähigkeit bleibt auch bei einer zusätzlichen Variation des Luftverhältnisses, der Restgasrate oder der Wassereinspritzungsrate bei verschiedenen Motorlasten und -drehzahlen bestehen. Somit bietet das Modell nicht nur die Möglichkeit der Potentialabschätzung CO_2-neutraler Kraftstoffe, sondern auch die Grundlage für die Entwicklung zukünftiger Motorkonzepte.

Als Basis für die Modellentwicklung wurde eine systematische Messdatenanalyse durchgeführt, um die Relevanz des bisher unklaren Kraftstoffeinflusses auf die Flammenfaltung unter motorischen Randbedingungen zu evaluieren. Dazu wurde neben der Brenndauer eine auf Basis des gemessenen Zylinderdruckverlaufs berechnete, turbulente Flammengeschwindigkeit herangezogen. Die Untersuchung beider Größen zeigte einen zu vernachlässigenden Kraftstoffeinfluss auf die Flammenfaltung, was im Gegensatz zu bereits bestehenden Ergebnissen der Grundlagenforschung steht und sich über deutlich abweichende Randbedingungen erklären lässt. Insbesondere der höhere Druck unter motorischen Bedingungen, zusammen mit der in der Literatur beschriebenen Druckabhängigkeit des Kraftstoffeinflusses auf die Flammenfaltung, spielt eine dominierende Rolle. Zudem ermöglichte diese Auswertemethodik die Bewertung und den Ausschluss eines möglichen Einflusses hydrodynamischer Flammeninstabilitäten auf die motorische Verbrennung. Ein Vergleich der auf Basis des Zylinderdrucks berechneten, turbulenten Flammengeschwindigkeit mit Literaturdaten eines optischen Motors bestätigte nicht nur die Validität der Methode, sondern lieferte auch erste Hinweise auf den Einfluss lokaler Turbulenzunterschiede.

Als Konsequenz der Messdatenanalyse genügt zur Abbildung des Kraftstoffeinflusses im Brennverlaufsmodell die Berücksichtigung der kraftstoffabhängigen laminaren Flammengeschwindigkeit sowie der laminaren Flammendicke. Dazu wurde ein Modell für die laminare Flammendicke entwickelt sowie ein bestehendes Modell der laminaren Flammengeschwindigkeit erweitert und an Ergebnisse reaktionskinetischer Berechnungen kraftstoffindividuell adaptiert. Da Messungen der laminaren Flammengeschwindigkeit unter motorisch relevanten Randbedingungen aufgrund des Auftretens zellulärer Flammenstrukturen nicht möglich sind, ist deren Bestimmung nur mittels Reaktionskinetikrechnungen möglich. Allerdings hängen diese von dem verwendeten Reaktionsmechanismus ab, was eine indirekte Plausibilisierung notwendig macht. Diese wurde im Zuge der Brennverlaufsmodellvalidierung durchgeführt. Im Gegensatz zu messungsbasierten Modellen, deren Belastbarkeit sich besonders bei hohen Restgasraten oder hohen Luftverhältnissen als unzureichend erwiesen hat, bieten Reaktionskinetikrechnungen den Vorteil, dass sich notwendige Extrapolationen in den motorischen Randbedingungsbereich vermeiden lassen.

Für die reaktionskinetische Berechnung von Benzin ist die Definition eines repräsentativen Ersatzkraftstoffes notwendig. Untersuchungen ergaben, dass ein Gemisch aus iso-Oktan und n-Heptan die laminare Flammengeschwindigkeit von Benzin unterschätzt, weswegen zusätzlich Toluol als Ersatzkraftstoffkomponente Anwendung fand. Darüber hinaus wurde eine Methode zur Berücksichtigung der detaillierten Zusammensetzung von Erdgas entwickelt. Während für hohe Methananteile die Betrachtung von reinem Methan für die Modellierung ausreicht, zeigt sich besonders bei höheren Ethananteilen, dass diese nicht vernachlässigt werden sollten. Ergänzend wurde Wasserstoff als zusätzliche Kraftstoffkomponente abgebildet, um den Effekt einer Beimischung, beispielsweise bei Verwendung von Netzgas als Kraftstoff, untersuchen zu können.

Die Qualität des Brennverlaufsmodells ließ sich weiter steigern, indem die Berechnungsgleichung für die turbulente Flammengeschwindigkeit ausgetauscht wurde. Unter Zuhilfenahme des *Borghi*-Diagramms und dessen der Literatur entnommenen Erweiterungen konnte gezeigt werden, dass der Motorbetrieb hauptsächlich im Bereich der gefalteten Flammen stattfindet und sich nicht in das Gebiet dünner Reaktionszonen erstreckt. Dadurch ist sichergestellt, dass verfügbare Berechnungsgleichung der turbulenten Flammengeschwindigkeit prinzipiell auch unter motorischen Randbedingungen anwendbar sind. Neben einem Vergleich basierend auf Literaturdaten wurde die zylinderdruckbasierte Berechnung der turbulenten Flammengeschwindigkeit erneut herangezogen, um die verlässlichste Berechnungsgleichung zu bestimmen. Dabei ließ sich ein unplausibler Druckeinfluss sowie eine deutliche Unterschätzung des Einflusses der laminaren auf die turbulente Flammengeschwindigkeit bei Verwendung der im Ausgangsmodell implementierten Berechnungsgleichung nach *Damköhler* beobachten. Besonders letztere Problematik hätte die Vorhersagegüte in Bezug auf eine Kraftstoffvariation beeinträchtigt, weswegen der *Damköhler*- durch den *Peters*-Ansatz ersetzt wurde.

Darüber hinaus wurden auch Unterschiede in der Definition des Referenzradius in der Bewertung der Berechnungsgleichungen berücksichtigt, um die Konsistenz des gesamten Brennverlaufsmodells sicherzustellen. In diesem Zusammenhang konnte gezeigt werden, dass im idealisierten Fall kein Kalibrierungsparameter im Modell notwendig ist, um gemessene Brennverläufe vorhersagen zu können. Der Rückschluss vom idealisierten auf den realen Fall erlaubt es somit, die Notwendigkeit vorhandener Abstimmparameter eindeutig zu begründen. Einer dieser Parameter ist die Position des Flammenzentrums, welche zur Berechnung der Flammenoberfläche auf Basis einer vereinfachten, zylindrischen Brennraumgeometrie verwendet wird. Es konnte gezeigt werden, dass durch Anpassung des Flammenzentrums mit der vereinfachten Geometrie ähnlich gute Ergebnisse wie mit der detaillierten Brennraumgeometrie erzielt werden können.

Eine systematische Analyse verschiedener Einflüsse auf die Vorhersagegüte des Brennverlaufsmodells zeigte die Notwendigkeit, den bisherigen Ansatz zur Berechnung der charakteristischen Brennzeit, einer Kerngröße von Brennverlaufsmodellen des Entrainment-Typs, weiterentwickeln zu müssen. Der neu entwickelte Berechnungsansatz ermöglicht eine aus phänomenologischer Sicht sinnvollere Abbildung von Flammenfaltungseffekten über alle Größenskalen des turbulenten Spektrums. Dadurch wird die Überschätzung der laminaren Flammengeschwindigkeit als Einflussgröße, wie sie im ursprünglichen Berechnungsansatzes auftrat, verhindert. Die mathematische Interpretation des Ansatzes, basierend auf der Analyse des Einflusses von Brennverläufen einzelner Arbeitsspiele auf den Brennverlauf eines mittleren Arbeitsspiels, liefert eine neue Erklärung, weswegen die Berechnung einer charakteristischen Brennzeit grundsätzlich notwendig ist. Damit wird auch die Kopplung des Brennverlaufs an ein bestimmtes Niveau der Zyklenschwankungen deutlich. Basierend darauf wurde die Theorie entwickelt, den Brennverlauf eines mittleren Arbeitsspiels über die Simulation und anschließende Mittelung der Brennverläufe einzelner Arbeitsspiele zu berechnen, anstatt die charakteristische Brennzeit zu verwenden. Dazu ist allerdings die genaue Kenntnis lokaler, zyklenschwankungsabhängiger Effekte notwendig.

Einen ersten Schritt in diese Richtung stellt die Neuentwicklung eines mathematischen Modells zur Berücksichtigung der lokalen Verteilung der turbulenten kinetischen Energie (TKE) dar. Als Entwicklungsgrundlage dienten 3D-CFD Daten, mit deren Hilfe neben der generellen Verteilung der TKE auch die Unabhängigkeit derselben von der Kolbenposition bei Referenzierung auf den relativen Flammenradius festgestellt werden konnte, wohingegen die Brennraumform einen signifikanten Einfluss aufwies. Dieses Modell verbessert bereits unter Verwendung der charakteristischen Brennzeit die Güte des Brennverlaufsmodells. Dies ist auf die geometrischen Verhältnisse zurückzuführen, welche bereits bei mittleren Massenumsätzen aufgrund der Expansion des verbrannten Luft-Kraftstoff-Gemisches, resultierend aus der Dichtereduktion durch Verbrennung, einen geringen Abstand zwischen Flamme und Zylinderwand verdeutlichen, womit die dort reduzierte TKE von Relevanz sein kann. Darüber hinaus konnte aus der lokalen Verteilung ein Anstieg der TKE bei kleinen Flammenradien ermittelt werden. Gemeinsam mit dem bereits beschriebenen Einfluss von Einzelarbeitsspielen auf den Brennverlauf eines mittleren Arbeitsspiels liefert diese Betrachtung eine neue Begründung für mögliche Modellabweichungen in der frühen Verbrennungsphase, welche bisher meist auf eine Zunahme des wirksamen Turbulenzspektrums zurückgeführt wurden. Die TKE-Verteilung ließ außerdem Rückschlüsse auf eine bereits in der Literatur beschriebene, ungleichförmige Flammenausbreitung, ebenfalls während der frühen Verbrennungsphase, zu, welche eine Verschiebung des Flammenmittelpunkts zur Folge hat und somit erneut dessen Verwendung als Abstimmparameter bestätigt.

Das weiterentwickelte Brennverlaufsmodell wurde anhand von sechs konzeptionell unterschiedlichen Motoren validiert. Diese decken ein breites Spektrum an Verdichtungsverhältnissen, Hub-Bohrungs-Verhältnissen sowie Turbulenzniveaus ab und umfassen ebenfalls einen Gasgroßmotor. Methan, CNG, Benzin, Ethanol, Methanol, Methylformiat und DMC+ wurden als Kraftstoffe verwendet. Durch die große Bandbreite an regulären sowie alternativen und synthetischen Kraftstoffen konnte die Güte des Modells in Bezug auf die Vorhersage des Kraftstoffeinflusses ohne Neuabstimmung bestätigt werden. Diese blieb auch bei einer Variation von Last, Drehzahl, Luftverhältnis, Restgasrate und Wassereinspritzungsrate erhalten. Die verlässliche Vorhersage des Lasteinflusses unterstreicht außerdem das Fehlen eines Einflusses hydrodynamsicher Flammeninstabilitäten nach *Darrieus-Landau*, welches bereits in der Messdatenanalyse beobachtet wurde. Die korrekte Abbildung der Drehzahlvariation zeugt von einer plausiblen Reaktion des Modells auf Änderungen der Turbulenz. Neben der Vorhersage des Kraftstoffeinflusses unterstreicht auch die gleichbleibende Vorhersagequalität bei einer Variation von Luftverhältnis und Restgasrate das Fehlen eines signifikanten Kraftstoffeinflusses auf die Flammenfaltung, da dieser in der Theorie über eine Restgasvariation konstant bleibt, sich aber deutlich mit dem Luftverhältnis verändert. Somit lässt sich der Kraftstoffeinfluss rein über die laminare Flammengeschwindigkeit sowie die laminare Flammendicke abbilden. Damit verspricht das Modell eine einfache Ergänzung zusätzlicher Kraftstoffe durch Erweiterung der entsprechenden Untermodelle auf Basis reaktionskinetischer Berechnungen. Im Rahmen der Modellvalidierung wurde auch der Einfluss verschiedener Reaktionsmechanismen untersucht. Dadurch konnte gezeigt werden, dass die Ergebnisse der gewählten Mechanismen plausibel sind, womit indirekt die Werte der reaktionskinetisch bestimmten, laminare Flammengeschwindigkeit bei motorischen Randbedingungen verifiziert wurden. Allerdings ist diese Erkenntnis an die gewählte Modellstruktur gekoppelt.

In einer anschließenden Untersuchung konnte bestätigt werden, dass der Kraftstoffeinfluss auch bei der Simulation des Motorklopfens allein durch den Austausch der Chemie abgebildet werden kann. Dazu wurde ein bestehendes Modell der Zündverzugszeit unter Verwendung reaktionskinetischer Berechnungen um die zuvor genannten Kraftstoffe erweitert. Das Modell ermöglichte ohne Neuabstimmung die Vorhersage des Klopfbeginns von Einzelarbeitsspielen bei verschiedenen Kraftstoffen. Diese Untersuchung konnte auch die auf Basis einer Messdatenanalyse gezogene Schlussfolgerung unterstützen, wonach ein Einfluss der Auslagerung von unverbrannter Gemischmasse aus dem Feuerstegbereich auf das Motorklopfen zu vernachlässigen ist: Die nach Spitzendruck auftretende Auslagerung könnte durch Abkühlungseffekte die Selbstzündung des unverbrannten Luft-Kraftstoff-Gemisches verzögern oder sogar unterbinden, jedoch wurde keiner dieser Effekte beobachtet. Dies unterstützt die bestehende Theorie, dass Klopfen in einer großen Endgastasche stattfindet. Auch die Vorhersage der Klopfgrenze auf Basis mittlerer Arbeitsspiele für verschiedene Kraftstoffe wurde validiert. Dabei konnte ein starker Einfluss der unterschiedlichen Zyklenschwankungen bei verschiedenen Kraftstoffen beobachtet werden. Da Informationen über das Zyklenschwankungsniveau bei Einzelarbeitsspielmittelung verloren gehen, ist die Vorhersage der Klopfgrenze an ein bestimmtes Niveau der Zyklenschwankungen gekoppelt. Ist das Zyklenschwankungsniveau jedoch vergleichbar, so ist auch die Vorhersage der kraftstoffabhängigen Klopfgrenze verlässlich.

Zusammenfassend ist es also möglich, auf Basis der Ergebnisse dieser Arbeit den Kraftstoffeinfluss sowohl auf die Verbrennung als auch auf das Klopfen für reguläre, alternative und synthetische Kraftstoffe in der 0D/1D-Simulation vorherzusagen. Außerdem versprechen die Modelle eine einfache Erweiterung um zusätzliche Kraftstoffe.

1 Introduction

Climate change demands extensive measures on a global scale to limit its impact on the environment to a tolerable level. In [4], for example, a necessary CO_2 emission reduction by 60 % compared to 1990 in the European Union was estimated to keep the global warming below 2 °C. Assuming an equal share of the necessary CO_2 reduction among all CO_2 emitters, the transport sector faces serious challenges. Further improvement of the already highly optimized spark-ignition combustion engine, for example by using a lean burn concept, enables its CO_2 emissions to be decreased, and thus can be employed as a medium-term measure. Since internal combustion engines are still present in scenarios for the year 2050 [12], for example in the heavy duty sector or for ship propulsion, such improvements are of great importance. However, engine optimization alone will not be sufficient, especially when aiming at a net CO_2 neutrality.

Apart from hybrid powertrains or battery electric vehicles (which also have to deal with a negative environmental impact like e.g. the use of rare earth elements, recycling and the high energy demand in battery production [110]), replacing gasoline (or diesel) with alternative fuels is a promising measure to further reduce CO_2 emissions. Fossil natural gas already allows a CO_2 emission reduction by 25 % [83] due to a higher hydrogen to carbon ratio alone. An even stronger reduction is possible when adapting the engine design to the fuel characteristics, for example by taking advantage of the higher knock resistance of natural gas. The synthetic production of fuels and their usage as pure fuels or blends with gasoline or methane, could reduce CO_2 emission even more, ultimately leading to CO_2 neutrality.

Recently, studies like [148] investigated the potential of different fuel candidates. In addition to production processes and toxicity, their influence on engine efficiency is of importance. In this context, engine simulation is a useful tool to compare different fuels, without the need for their actual availability. Due to a fast model set-up and low computation times, 0D/1D simulation promises to be very efficient, even allowing the evaluation of real driving scenarios.

© The Author(s), under exclusive license to
Springer Fachmedien Wiesbaden GmbH, part of Springer Nature 2021
S. Hann, *A Quasi-Dimensional SI Burn Rate Model for Carbon-Neutral Fuels*,
Wissenschaftliche Reihe Fahrzeugtechnik Universität Stuttgart,

Concerning engine knock and the influence of direct fuel injection on turbulence, various research projects [50][42][137] were performed recently. In order to evaluate the fuel influence on engine efficiency and to optimize the engine concept for a specific fuel, a reliable burn rate model is also necessary. Therefore, the main objective of this work is to develop a burn rate model with the ability to predict the influence of fuel changes on engine operation with a constant set of calibration parameters. Furthermore, the model should cover a wide range of boundary conditions, such as lean mixtures and high rates of EGR or water injection, in order to enable the development of future engine concepts. A high predictive ability would also allow the model to be calibrated on the basis of measurement data using a standard fuel, consequently increasing the reliability of subsequent simulations of alternative fuels. With that, valuable engine test bed resources could be used more efficiently, for example by verifying the simulation results directly on an improved engine concept instead of the base engine.

All relevant influences of the fuel on the combustion process need to be accounted for in order to achieve a high level of predictive ability. To illustrate the current influence of the fuel and describe ways to model other relevant influences, such as turbulence, the basics of quasi-dimensional burn rate simulation are described in Chapter 2. Subsequently, results of the fundamental research concerning laminar and turbulent flame propagation are given. This includes the fuel influence on flame wrinkling, as observed in turbulent flame speed measurements. Furthermore, subjects such as different definitions of the turbulent flame speed are also discussed, since they have an influence on the model development. However, due to a difference in boundary conditions, the applicability of fundamental research results to engine operation is uncertain. The discrepancy between available measurements of laminar flame speeds and engine-relevant boundary conditions is one example highlighting this fact. Thus, it is reasonable to compare engine measurement data with expected combustion characteristics from fundamental research to evaluate their relevance.

In Chapter 3, such an analysis of engine measurements is performed for a variation of air-fuel ratios, using methane, ethanol and gasoline as fuels. From fundamental research, those fuels are expected to show significant differences in flame wrinkling, which also change with varying air-fuel ratios. In measurement data, none of these influences could be observed. With that, the main fuel influence on combustion characteristics seems to be linked to changes in the laminar flame speed alone.

Following these findings, an entrainment-type burn rate model is improved in Chapter 4. The improvement process includes the development of models for the laminar flame speed and the laminar flame thickness, for which reaction kinetics calculations are used as a basis. Methane, CNG substitutes, methanol, ethanol, gasoline, hydrogen, methyl formate and DMC+ (65 vol–% dimethyl carbonate, 35 vol–% methyl formate) are considered as possible fuels. Furthermore, different turbulent flame speed models are evaluated. The calculation of the characteristic burn-up time as one of the key features of the entrainment approach is adapted and interpreted phenomenologically and mathematically. The latter highlights the influence of single working cycles on the burn rate of averaged working cycles and investigates the need for a distinct model to cover the early phase of combustion. The mathematical interpretation furthermore underlines the linking of the simulated burn rate to a fixed level of cycle-to-cycle variations (CCV). An existing CCV model is then adapted to the improved burn rate model, enabling the investigation of CCV influences on fuel consumption and engine operation limits. Additionally, a 3D CFD-based mathematical model is implemented in the burn rate model to cover the influence of the local turbulence distribution in the combustion chamber.

In Chapter 5, the improved burn rate model is validated on six different engines for various fuels and operating conditions. The specifications of the engines include a high turbulence concept, a long stroke configuration, a wide range of compression ratios and a large gas engine. The extensive validation promises a high predictive ability of the burn rate model not only for a fuel variation, but also for lean engine operation and high rates of EGR or water injection. This allows the potential of different standard and alternative or synthetic fuels to be evaluated, as well as optimizing the engine design, without the need to recalibrate the burn rate model.

Going beyond the main scope of this thesis, Chapter 6 presents a brief investigation of the fuel influence on engine knock. For the prediction of fuel-dependent burn rate changes, it was sufficient to model the laminar flame speed changes. This, to a certain extent, represents the chemical effect of the fuel.

For engine knock, the approach translates to modeling the fuel-dependent ignition delay time in order to predict engine knock. Therefore, after expanding
existing ignition delay time models for gasoline and CNG to also cover ethanol
and toluene, the predicted knock onset of single and averaged working cycles
is evaluated. Furthermore, the influence of top land gas discharge on engine
knock is investigated by means of a measurement data analysis as well as a
comparison of simulated and measured knock onsets of single working cycles.

During the course of this work, intermediate results were already published in
[71], [72] and [74], where [74] is based on the Master's thesis of the author
[70]. These publications served as basis for the present thesis. Paragraphs
taken from [71] and [74] are marked individually throughout this document.
Chapter 2.3.2, Chapter 2.4.1 and large parts of Chapters 2.5 to 6 were taken
from [72].

2 Fundamentals

In the context of reducing CO_2 emissions, alternative or synthetic fuels can significantly improve the ecological footprint of internal combustion engines. To adapt an engine to changing fuel characteristics, 0D/1D simulations offer an efficient way to streamline the engine development process. A prerequisite for this is the reliable modeling of the fuel influence on combustion. This chapter presents common features of the most popular burn rate models and gives a more detailed description of an example burn rate model, including the fuel influence therein. Furthermore, the fundamentals of laminar and turbulent flame speeds as well as the fuel influence on flame wrinkling are discussed, since they are important inputs for the burn rate model.

2.1 Common Features of Phenomenological SI Burn Rate Models

In [31], *Demesoukas et al.* gave an overview of the most common burn rate models of SI engines for application in 0D/1D engine simulation. These are the entrainment or eddy burn-up model published in [14], the fractal model (see e.g. [17]) and the flame surface density model (see e.g. [129]). All models assume a hemispheric flame propagation, starting at the spark plug. The flame surface area is calculated as the intersection of a sphere and the combustion chamber. Furthermore, all models assume a turbulent flame speed for the propagation of the flame surface. In the fractal and the flame surface density model, this speed is calculated as the product of the laminar flame speed and a wrinkling factor. The wrinkling factor accounts for the flame surface increase due to turbulence-induced flame wrinkling. Both models only differ in the way of calculating the wrinkling factor. In the entrainment model, the turbulent flame speed is calculated, for example, as the sum of the laminar flame speed and the turbulent fluctuation velocity. Again, the influence of turbulence increases the flame propagation speed. To calculate the burn rate, the entrain-

Springer Fachmedien Wiesbaden GmbH, part of Springer Nature 2021
S. Hann, *A Quasi-Dimensional SI Burn Rate Model for Carbon-Neutral Fuels*,
Wissenschaftliche Reihe Fahrzeugtechnik Universität Stuttgart,

ment model furthermore employs a characteristic burn-up time (see eq. 2.2). In the fractal and the flame surface density model, the burn rate is directly calculated as the product of the unburnt density, the flame surface and the turbulent flame speed. As will be shown in Chapter 2.4.1, the different model approaches correspond to different definitions of the turbulent flame speed, ultimately justifying the characteristic burn-up time approach in the entrainment model.

When comparing the physical inputs used to calculate the burn rate (as summarized in [31]), all models rely on the same main parameters: related to turbulence, those are the turbulent fluctuation velocity and the integral length scale. The influence of fuel and boundary conditions like pressure or temperature are accounted for by using the laminar flame speed. This underlines a certain similarity of all the models, since the laminar flame speed is increased by the influences of turbulence in every single one.

2.2 Example Burn Rate Model

To illustrate the details of burn rate modeling, the main model features of a quasi-dimensional, entrainment-type burn rate model for SI engines are described in the following. As the basic model structure of this type is widely used in commercial software, it is a useful example. The model was originally developed in [14] and [133] and was described in great detail in [60], [62] and [61]. The general validity of this approach will be proven in the course of this thesis.

[147] In the theory of the entrainment approach, the flame front divides the combustion chamber into a burnt and an unburnt zone. The flame front itself is not considered as an additional, thermodynamical zone (see Figure 2.1). Therefore, as stated in [61], the burn rate model is based on a two-zone working process calculation, where the calorific properties are calculated for user-definable fuels using the approach published in [63] and [65]. The equations describing the thermodynamic relationship between both zones are elaborated in detail in [140].

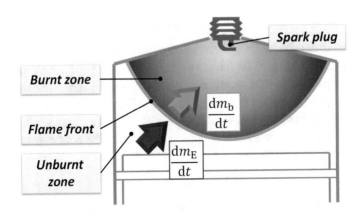

Figure 2.1: Schematic representation of the entrainment approach [147]

The key characteristic of the entrainment model is the calculation of the burn rate $\mathrm{d}m_\mathrm{b}$ on the basis of the characteristic burn-up time τ_L and the mass m_F entrained into the flame front (eq. 2.1).

$$\frac{\mathrm{d}m_\mathrm{b}}{\mathrm{d}t} = \frac{m_\mathrm{F}}{\tau_\mathrm{L}} = \frac{m_\mathrm{E} - m_\mathrm{b}}{\tau_\mathrm{L}} \qquad\qquad \text{eq. 2.1}$$

As shown in eq. 2.2, the characteristic burn-up time is defined as the ratio of the *Taylor* microscale l_T to the laminar flame speed s_L.

$$\tau_\mathrm{L} = \frac{l_\mathrm{T}}{s_\mathrm{L}} \qquad\qquad \text{eq. 2.2}$$

The *Taylor* microscale depends on the integral length scale l_int, the turbulent fluctuation velocity u', the turbulent kinetic viscosity ν_T and the *Taylor* factor χ_Taylor. The value of the latter is assumed to be 15 [135] (eq. 2.3).

$$l_\mathrm{T} = \sqrt{\chi_\mathrm{Taylor} \cdot \frac{\nu_\mathrm{T} \cdot l_\mathrm{int}}{u'}} \qquad\qquad \text{eq. 2.3}$$

To close the model, the entrainment mass flow into the flame front ($\mathrm{d}m_\mathrm{E}$) needs to be calculated. For this, eq. 2.4 and eq. 2.5 are used.

$$\frac{\mathrm{d}m_\mathrm{E}}{\mathrm{d}t} = \rho_\mathrm{u} \cdot A_\mathrm{fl} \cdot u_\mathrm{E} \qquad\qquad \text{eq. 2.4}$$

$$u_E = u' + s_L \qquad \text{eq. 2.5}$$

[72] The gas density in the unburnt zone (ρ_u) is calculated with the corresponding mass and volume. The flame surface A_{fl} is calculated according to [62]. There, a hemispherical flame propagation in a cylindrical combustion chamber was used to calculate the free flame surface and the contact area between the burnt zone and the combustion chamber walls. The spark plug position can be defined freely and serves as a calibration parameter. The heat release rate dQ_b is finally calculated using eq. 2.6.

$$dQ_b = dm_b \cdot H_{u,\text{mix,grav.}} \qquad \text{eq. 2.6}$$

To calculate the turbulent fluctuation velocity u', two different k-ε turbulence models are available: a homogeneous, isotropic model published in [62] and a quasi-dimensional model, introduced in [15]. The latter model simplifies the tumble flow structure using a *Taylor-Green* eddy [134]. The resulting flow field yields the turbulence production using the same k-ε sub-models as 3D CFD software, but with simplified combustion chamber geometries. Thus, influences on the turbulent kinetic energy (TKE) – such as changing cam durations and timings – can be predicted. In [46], the latter model was expanded to cover the influence of swirl and squish flow, combined swirl-tumble charge motion (e.g. due to asymmetric valve lift) and direct fuel injection on turbulence. In this study, the option to import the TKE and integral length scale from 3D CFD calculations was added to the simulation code. As will be described in Chapter 4.5, a method to account for the local distribution of the TKE was developed as well.

For the laminar flame speed s_L, the approaches of *Gülder* [68] (eq. 2.8) and *Heywood* [75] (eq. 2.9) are used in the baseline burn rate model for methane and gasoline, respectively. Both models are based on eq. 2.7. They were fitted to measurements of laminar flame speeds. As will be elaborated in more detail in Chapter 2.3.1 and Chapter 2.3.2, those measurements are limited to boundary conditions irrelevant for engine operation. This leads to a qualitatively and especially quantitatively incorrect change in the laminar flame speed with boundary conditions, in comparison with reaction kinetics calculations (see Chapter 4.1).

$$s_L(\Phi, T_u, p, Y_{EGR}) = s_{L,0}(\Phi) \cdot \left(\frac{T_u}{T_{\text{ref}}}\right)^{\alpha} \cdot \left(\frac{p}{p_{\text{ref}}}\right)^{\beta} \cdot (1 - F \cdot Y_{EGR}^n) \qquad \text{eq. 2.7}$$

$$s_{L,0}(\Phi) = 0.422 \cdot \Phi^{0.15} \cdot e^{-5.18 \cdot (\Phi - 1.075)^2}$$

$$\alpha = 2$$

$$\beta = -0.5 \qquad\qquad \text{eq. 2.8}$$

$$F = 2.5$$

$$n = 1$$

$$s_{L,0}(\Phi) = 0.305 - 0.549 \cdot (\Phi - 1.21)^2$$

$$\alpha = 2.18 - 0.8 \cdot (\Phi - 1)$$

$$\beta = -0.16 + 0.22 \cdot (\Phi - 1)$$

$$F = 2.06 \qquad\qquad \text{eq. 2.9}$$

$$n = 0.77$$

2.3 Laminar Flame Speed

The laminar flame speed s_L describes the propagation speed of an unstretched flame front in a quiescent gas mixture of fuel and oxidizer. As schematically illustrated in Figure 2.2, the flame front can be divided into three different zones: the preheat zone, the inner layer or reaction zone and the oxidation layer. In the preheat zone, the unburnt temperature T_u is increased by heat conduction. The fuel concentration, in this example methane, decreases due to species diffusion into the reaction zone, driven by concentration gradients. The reaction zone is dominated by rapid chemical processes, where the fuel molecules are broken down and many intermediate species are formed [136]. In the oxidation layer, CO is oxidized to CO_2 by slower chemical processes, resulting in the greatest heat release across the flame front. From the gradients of temperature and species concentrations, it is apparent that the flame propagation is dependent on heat and mass diffusion, besides chemical processes. More details were given in [136] and [119]. The heat release causes the temperature to rise to its maximum value, the burnt temperature T_b. The temperature increase decreases the density, since the pressure is assumed to be constant across the flame front [136]. This is justified by the much higher speed of sound compared to the laminar flame speed. The density decrease leads to an increase in flame speed on the burnt side of the flame front. Therefore, the laminar flame speed can be related either to the burnt or the unburnt side of the flame front. In this thesis, the laminar flame speed is always related to the unburnt side.

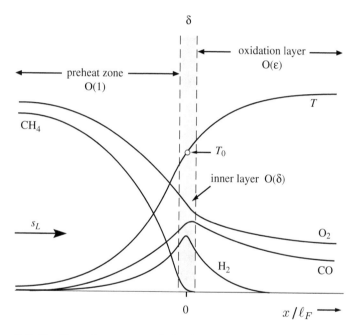

Figure 2.2: Profile and structure of a laminar flame (edited version from [119], original source: [118])

The value of the laminar flame speed is dependent on boundary conditions such as unburnt temperature, pressure, fuel, air-fuel ratio, EGR rate and water injection rate. As it is an important input parameter of the burn rate model, ways of obtaining values for the laminar flame speed are described in the following.

2.3.1 Measurement of Laminar Flame Speeds

Laminar flame speeds can be measured in an explosion (or combustion) bomb, for example. Other methods are the Bunsen burner (description in [119]) or the heat flux burner (example application in e.g. [34]). As the combustion bomb method is widely used and turbulent flame speeds are measured in similar devices, its fundamentals are reproduced here. An exemplary application of this method for laminar and turbulent flames can be found in [112].

The explosion bomb consists of an optically accessible combustion chamber filled with gaseous, premixed fuel and oxidizer. This mixture is ignited by electrodes in the center of the chamber, followed by a spherical flame propagation. The flame front propagation is tracked via e.g. schlieren photography through the chamber window. The tracking is not only limited by the size of the window, but also by the maximum acceptable pressure rise in the chamber due to combustion. The stretched laminar flame speed s_f results from the flame front radius progression with time, see eq. 2.10. The density ratio of burnt and unburnt gas (ρ_b/ρ_u) accounts for the thermal expansion of the burnt gas due to combustion.

$$s_f = \frac{dR}{dt} \cdot \frac{\rho_b}{\rho_u} \qquad\qquad \text{eq. 2.10}$$

The flame stretch rate is a result of the spherical flame geometry and can be calculated as

$$\kappa \overset{\text{general}}{=} \frac{1}{A} \cdot \frac{dA}{dt} \overset{\text{spherical}}{=} \frac{2}{R} \cdot \frac{dR}{dt}. \qquad\qquad \text{eq. 2.11}$$

When plotting s_f over κ, Figure 2.3 is obtained. The figure illustrates the change in the stretched laminar flame speed with stretch rate. By linearly extrapolating to zero stretch rate, the unstretched laminar flame speed s_L is calculated. The slope of the extrapolation line is the *Markstein* length and describes the reaction of the flame to stretch. More details will be given in Chapter 2.5. As Figure 2.3 shows, the measurement of the stretched laminar flame speed is affected by ignition and the onset of cellularity. This can lead to uncertainties in the extrapolation and consequently the unstretched laminar flame speed value. The cellularity can not only be caused by *Markstein* or *Lewis* effects (see Chapter 2.5), but also by the *Darrieus-Landau* (DL) instability. As the DL instability not only influences laminar flame speed measurements, but possibly also the turbulent flame speed, it will be described in more detail in Chapter 2.3.2. According to *Law et al.* [89], measurements in explosion bombs can also be affected by the buoyantly unstable nature of the spherical flame configuration: the burnt gas is surrounded by the denser, unburnt gas and can thus be compared to an air bubble in water. All the mentioned uncertainties in combination with sensor accuracy cause a measurement error of about 5 to 10 % [34] [80].

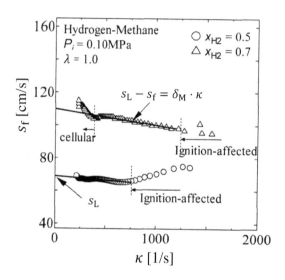

Figure 2.3: Measurement of s_L and *Markstein* length δ_M [112]

2.3.2 The *Darrieus-Landau* Instability

The hydrodynamic instability of flames, first described by *Darrieus* [30] and *Landau* [88], was explained in more detail in [89]. It is relevant for flame deformations which are significantly larger than the laminar flame thickness δ_L. Thus, the flame thickness can be neglected. To derive the origin of the instability, local speed changes and conservations are considered (see Figure 2.4): at a flame deformation, a flame speed normal to the flame front (u^0) and one tangential to the local flame front (u_t) exist. The flame speed normal to the flame is increased in the burnt zone due to thermal expansion. The flame speed tangential to the local flame front is not affected. At a convex flame deformation (towards the unburnt), these relations lead to a widening of the local stream tube of unburnt gas (top part of Figure 2.4). With that, the local

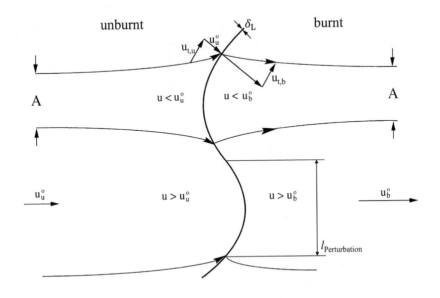

Figure 2.4: Schematic explanation of the *Darrieus-Landau* instability [89]

stream velocity is reduced ($u < u_u^0$). Since the local flame speed remains constant, the relative velocity between unburnt stream and flame is increased. As a result, the convexity grows. At a concavity, the opposite is true. With increasing pressure and temperature or decreasing λ or EGR rate, the laminar flame thickness decreases. With that, a flame disturbance of constant size is more likely to be increased. This not only limits the measurement of laminar flame speeds to relatively low temperatures and pressures, but might also increase flame wrinkling within the engine. In Figure 2.10, the area of DL instability influence is highlighted in the *Borghi* diagram. Considering the trajectory of an engine in the *Borghi* diagram (see Figure 2.10 and [25]), the influence of the DL instability on engine combustion should be small to non-existent. However, in a different publication [150], the area of DL instability influence was considered to be significantly larger and, with that, engine-relevant. This uncertainty was investigated using engine measurement data, as will be described in Chapter 3.1.

The enhancement of flame disturbances due to the DL instability restricts the available measurement data to low temperatures and pressures. As highlighted in Figure 2.5, much higher values are reached during the high pressure part of an engine cycle. Consequently, only reaction kinetics calculations can provide values of the laminar flame speed at engine-relevant boundary conditions.

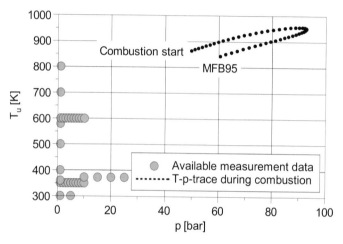

Figure 2.5: Available measurement data from [80][34][16][22][139][76][35] in comparison with a temperature and pressure trace of an engine (high pressure cycle) [71]

2.3.3 Reaction Kinetics Calculations

[136] Reaction mechanisms are the basis of reaction kinetics calculations. Such mechanisms are developed for specific fuels and contain the equations of all (known) elementary reactions taking place during the combustion. Their reaction rate coefficients are studied in detail for a wide range of boundary conditions to investigate their temperature and pressure dependency. Since elementary reactions follow known, fundamental principles, it is reasonable to use a reaction mechanism beyond its range of validation, which is based on measured laminar flame speeds and ignition delay times. Reaction mechanisms also contain the heat and mass transfer properties as well as thermodynamic properties of all (relevant) molecules. [74]

[74] To perform reaction kinetics calculations, a software like Cantera [58] is needed to utilize the information stored in a reaction mechanism. Cantera offers one-dimensional flames as a calculation scenario (compare Figure 2.2), which is used to determine laminar flame speeds and laminar flame thicknesses. The equations to be solved in such a scenario were described in great detail in [136]. Via Python or Matlab, calculations in Cantera can be automated and thus be performed for a wide range of boundary conditions with low effort. This method was employed to generate a database of laminar flame speeds and to quantify the influence of different boundary conditions. The database was used as a reference in the development process of a laminar flame speed model (see Chapter 4.1). Furthermore, the calculation results help to gain a general understanding of possible reasons for the boundary condition influences, in addition to fundamental research results taken from the literature. This general understanding not only helps to estimate the influence of engine operation conditions on the laminar flame speed, but also allows for a first comparison of different engine concepts, as highlighted in the following.

[74] Figure 2.6 exemplarily illustrates the calculated laminar flame speeds of methane for two water injection rates Y_{H2O} at different temperatures and pressures. Higher temperatures increase the speed of chemical reactions by increasing the probability of molecule collisions (collision theory, [136]). With that, the laminar flame speed increases with temperature. In contrast, an increase in pressure decreases the laminar flame speed. This influence can be explained by *Le Chatelier's* principle [136]: high pressures promote reactions with fewer products than reactants. With a lower number of molecules, the pressure is reduced. For combustion reactions, this can be translated to the acceleration of chain-breaking reactions. In these reactions, fuel radicals recombine to a single stable molecule, hence breaking the chain reaction of the combustion and consequently decreasing the flame speed. The influence of water injection was investigated in great detail by *Dlugogorski et al.* [36]. They identified mixture dilution and heat sink effects of the additional water as major reasons for the decrease of the laminar flame speed, while chemical effects were considered negligible.

[71] Figure 2.7 shows, besides the difference in flame speed of different fuels and reaction mechanisms, the influences of λ. With increasing amounts of additional air at lean mixtures or fuel at rich mixtures, respectively, the flame

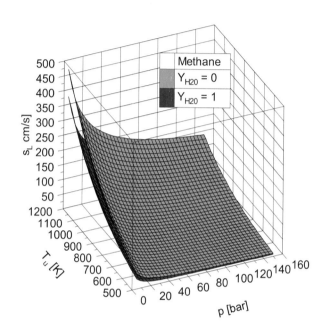

Figure 2.6: Laminar flame speed s_L for different pressures, temperatures and water injection rates, fuel: methane

speed decreases. This behavior is a result of two effects: first, the additional components dilute the mixture, which decreases the concentration of oxidant (rich) or fuel (lean). Second, the additional components work as heat sinks. Both effects reduce the speed of chemical reactions, and, therefore, the laminar flame speed. Figure 2.7 also shows a peaking of the laminar flame speed for slightly rich conditions. As elaborated in detail in [91], this is mainly the result of dissociation effects. When H_2O or CO_2 dissociate, one H_2 or CO molecule and only half an O_2 molecule are formed. In rich combustion products, H_2 and CO are present in larger amounts compared to lean combustion products that contain unburnt O_2. Due to the mole number differences in the dissociation products, lean mixtures are faced with less O_2 than rich mixtures are with H_2 and CO. Therefore, dissociation occurs to a larger extend in lean mixtures. Due to the endothermic character of dissociation reactions, this leads to a stronger reduction of the flame speed at lean mixtures, causing a peak for rich mixtures.

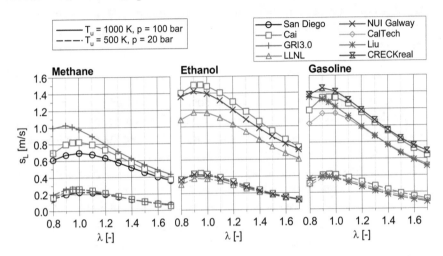

Figure 2.7: Influence of different reaction mechanisms on the laminar flame speed of different fuels at different values of λ [72]. Sources: SanDiego [102], Cai [22], GRI3.0 [49], LLNL [97], NUI Galway [105], CalTech [13], Liu [114], CRECKreal [125]

It is important to highlight the divergence of calculation results when using different reaction mechanisms, as depicted in Figure 2.7 for different fuels. A possible reason for the divergence is the limited validation data at higher temperatures and pressures. Furthermore, different reaction mechanisms are validated against different sets of measurement data, which are additionally influenced by measurement uncertainty. General progress in fundamental research also provides a more reliable foundation for the (further) development of reaction mechanisms, causing differences between older and more recent ones. Although all mechanisms shown in Figure 2.7 exhibit a similar reaction to changes in boundary conditions, their influence on the burn rate modeling needs to be taken into account. This will be described in Chapter 5.1.

[71] Figure 2.7 also illustrates the higher laminar flame speed of ethanol compared to gasoline. This example is used to give a general idea of possible reasons for the different laminar flame speed of different fuels. The flame propaga-

tion analysis of *Zeldovich* [151] served as a starting point for this investigation: based on several simplification, eq. 2.12 results from an eigenvalue-analysis of conservation equations for mass and energy, where α is the thermal diffusivity, v_L is the flame velocity and τ is the characteristic time of a single-step reaction.

$$v_L = \sqrt{\frac{\alpha}{\tau}} \qquad\qquad \text{eq. 2.12}$$

The equation illustrates that the flame speed is increased by either a higher thermal diffusivity or a higher chemical reactivity leading to a shorter chemical time scale. Both can be estimated from reaction kinetics calculations. In this example, the temperature gradient of the flame front was similar for both fuels. As the thermal diffusivity was also similar, the reason for the higher flame speed of ethanol could be of chemical nature: Appendix A2.1 compares the twelve most sensitive elementary reactions of ethanol and gasoline. For the latter, a toluene reference fuel (TRF) was employed as surrogate. These reactions were identified by performing a sensitivity analysis. Their numbering was taken from the *Cai et al.* mechanism published in [22], which was used for this investigation. The analysis was performed for $\lambda = 1$. Despite a large similarity between the sensitive reactions of ethanol and gasoline, a slight shifting of reaction paths for ethanol could be observed. For example, reaction no. 16 ($H + HO_2 \leftrightarrow 2\,OH$) showed a higher sensitivity for ethanol. Its forward rate coefficient is the highest of all reactions listed in Appendix A2.1, which explains its relatively low sensitivity. This reaction seems to offer an alternative and faster way of OH radical creation compared to the quite slow reaction no. 0. Since the OH molecule is the most important radical concerning the fuel breakdown [136], a faster OH creation shortens the chemical time scale τ in eq. 2.12 and thus increases the flame speed.

[71] Figure 2.8 compares the influence of residual gas (EGR) and λ on s_L for pure methane at a pressure of 1 bar and different temperatures. The observed behavior is also valid at pressures of 100 bar. The x-axes of Figure 2.8 are scaled to match the degree of mass dilution of residual gas and λ. As discussed above, the general reason for the decrease in laminar flame speed at increasing values of λ stems from mass dilution and heat sink effects due to the excess air. Both effects also apply to increasing EGR rates. However, at the same degree of mass dilution, the laminar flame speed is reduced more strongly by

residual gas. On the one hand, this results from different heat capacities of the air and the residual gas (diatomic versus triatomic molecules), which influence the heating of the unburnt gas as well as the flame temperature. On the other hand, the different chemical behavior of the reactive excessive air changes the reactions that take place during the combustion, compared to the nearly inert residual gas. Furthermore, in accordance with *Le Chatelier's* principle and the reason for off-stoichiometric s_L-peaking described in [91] (see above), the additional O_2 at lean mixtures inhibits the dissociation of CO_2 and H_2O. EGR, however, increases the concentration of CO_2 and H_2O, which supports the dissociation. Due to the endothermic character of dissociation, EGR decreases the temperature and, with that, the laminar flame speed stronger than lean mixtures. Figure 2.8 also highlights the influence of temperature on the relative change in the laminar flame speed, which needs to be taken into consideration when developing a model for the laminar flame speeds.

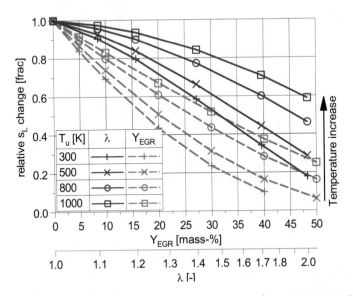

Figure 2.8: Relative laminar flame speed change with λ and EGR rate [71]

[71] Besides their importance for the modeling of laminar flame speeds, the differences between the influence of EGR and λ shown in Figure 2.8 give a

first indication of differences between high EGR and lean burn engine concepts. In part load, de-throttling can be achieved with both concepts. However, for the same amount of additional mass, an increase in λ reduces the laminar flame speed much less compared to EGR, which results in a faster and more stable combustion. At high loads, the reduction in peak temperature is a benefit of those concepts, since lower temperatures reduce the engine knock tendency, consequently allowing earlier spark timings. To evaluate both concepts in terms of their potential to reduce temperature, Appendix A2.2 depicts a derivative of Figure 2.8, where the additional mass is multiplied with the temperature-dependent value of the specific isochoric heat capacity c_v. For an isochoric change of state, a higher value of this product results in a lower temperature increase for the same amount of total heat input. As highlighted in Appendix A2.2, the laminar flame speed is again reduced much less for a lean burn engine concept, although the difference between both concepts is smaller compared to Figure 2.8. Based on this investigation, lean combustion seems to be advantageous. In Chapter 5.3, this conclusion – drawn from the laminar flame speed comparison – will be confirmed by engine test bed measurements.

2.3.4 Flammability Limit

The flammability limit describes the boundaries of mixture composition beyond which a self-sustaining flame propagation is no longer possible. These limits are also dependent on temperature and pressure. According to [79], no exact theory to explain the existence of those limits is available. One possible theory was proposed by *Law et al.* in [90]. Their simulative investigation showed a combined influence of radiative heat loss and chain reaction termination causing the flame propagation to no longer be self-sustained.

In [116], calculated laminar flame speeds were compared to measured flammability limits taken from [78]. This comparison related the flammability limit to a flame speed between 2.5 and 10 cm/s, as the change in laminar flame speed with temperature and mixture composition was similar to the change in the flammability limit. This range of laminar flame speeds can be used as a reference to identify relevant boundary conditions for the development and evaluation of a laminar flame speed model, see Chapter 4.1. Reaction kinetics calculations performed by the author supported the given range of laminar flame speeds. Furthermore, the reaction kinetics calculations gave a value for the laminar flame speed even beyond the measured flammability limit. As this indicates that the relevant influences causing the flammability limit are not accounted for in reaction kinetics calculations (or the measured flammability limit is influenced by effects specific for the measurement device), very low flame speed values taken from reaction kinetics have to be handled with care.

In engine combustion, a low laminar flame speed causes an unstable combustion, consequently increasing the cycle-to-cycle variations (CCV) [147]. For each engine, an individual limit for an acceptable level of the CCV is defined by the manufacturer and related to the coefficient of variation of the indicated mean effective pressure (COV_{IMEP}). Usually, this limit is at about 3 % COV_{IMEP}. In engine simulation, the laminar flame speed at the COV_{IMEP} limit can be estimated. This allows the relevance of the flammability limit in engine combustion at the engine operation limit to be evaluated, as will be done in Chapter 4.1.5.

2.4 Turbulent Flame Speed

The turbulent flame speed describes the propagation speed of a wrinkled flame front in a turbulent gas mixture of fuel and oxidizer. Its value is dependent on boundary conditions such as turbulence characteristics (turbulent fluctuation velocity u' and integral length scale l_{int}) and fuel (laminar flame speed s_L and laminar flame thickness δ_L). Like the laminar flame speed, it is an important input value of the burn rate model. Hence, relevant fundamentals are described in the following.

2.4.1 Measurement of Turbulent Flame Speeds

Like laminar flame speeds, premixed turbulent flame speeds are measured in spherical combustion bombs. Fans are used to generate an isotropic turbulence within the bomb. The turbulence level is controlled by changing the fan speed. More details were given in [18]. After the spherical flame propagation is initiated by a central ignition, the flame front radius is tracked over time. The turbulent flame speed is then calculated by using eq. 2.13.

$$s_\text{T} = \frac{\text{d}R}{\text{d}t} \cdot \frac{\rho_\text{b}}{\rho_\text{u}} \qquad \text{eq. 2.13}$$

The ratio of burnt to unburnt density accounts for the gas expansion of the burnt mixture caused by a temperature increase due to combustion. The turbulence leads to a wrinkling of the flame front. As was elaborated in more detail by *Bradley et al.* in [18], the choice of an unwrinkled reference radius R to be used in eq. 2.13 is arbitrary and significantly influences the value of s_T. The choice of a radius is limited by the leading edge (R_t in Figure 2.9) and the trailing edge (R_r in Figure 2.9) of the wrinkled flame front. Usually, either the leading edge or a mean location (R_j in Figure 2.9) is chosen as reference.

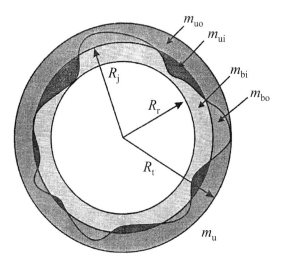

Figure 2.9: Schematic measurement of turbulent flame speeds and radius definition [18]

For the tip speed definition R_t, significant amounts of unburnt mass are present behind the flame front (m_ui). With that, the heat release rate is delayed compared to flame propagation. When using R_j as a reference, the amount of burnt mass outside of the radius (m_bo) and the amount of unburnt mass inside (m_ui)

are roughly similar and the flame propagation is in accordance with the heat release rate. This needs to be considered when comparing different turbulent flame speed models or flame speed measurements. Unfortunately, the chosen reference radius is not always given in detail in the literature.

The measurement technique used for flame front tracking (like e.g. schlieren photography or Mie scattering) also influences the value of s_T, since different positions in the flame front are visualized by different methods. As a result, the comparison of flame speed values from different studies is difficult. However, in several publications like [152], [26] and [106], a change in turbulent flame speed with changing fuel was observed. It was attempted to keep other boundary conditions, such as the laminar flame speed, as constant as possible. This change in turbulent flame speed is attributed to a fuel influence on flame wrinkling, usually described by the *Lewis* or *Markstein* number. Both concepts will be described in Chapter 2.5.

2.4.2 Regimes of Turbulent Combustion

The different regimes of turbulent combustion are described in the *Borghi* diagram, depicted in Figure 2.10. As the modeling of the turbulent flame speed is not only linked to a certain turbulent flame speed definition (see Chapter 2.4.1), but also to a certain regime of turbulent combustion, the classification of the engine-relevant regime is important for the evaluation of different turbulent flame speed models.

The x-axis of the *Borghi* diagram denotes the ratio of the integral length scale l_{int} to the laminar flame thickness δ_L. The y-axis denotes the ratio of the turbulent fluctuation velocity u' to the laminar flame speed s_L. The three main regimes (number 1 to 3 in Figure 2.10) are separated by certain values of the *Karlovitz* number $Ka = \tau_F/\tau_K$. $\tau_F = l_{int}/s_L$ is the chemical time scale and τ_K is the *Kolmogorov* time scale.

[127] For $Ka < 1$, the chemical processes are faster than the smallest turbulent time scale. In consequence, turbulent eddies only wrinkle the flame front, but cannot penetrate it. For $Ka < 100$, the turbulent time scale is small enough so that the smallest eddies can penetrate the preheat zone, but not the reaction zone of the flame (see Figure 2.2). The reaction zone is thus wrinkled by the

eddies, while the interaction of turbulent eddies with the preheat zone supports diffusive effects and sustains the increase in turbulent flame speed with turbulence. For $Ka > 100$, the smallest turbulent eddies can penetrate the reaction zone and disturb the chemical processes in the flame front. This can lead to local flame quenching, causing a broken reaction zone.

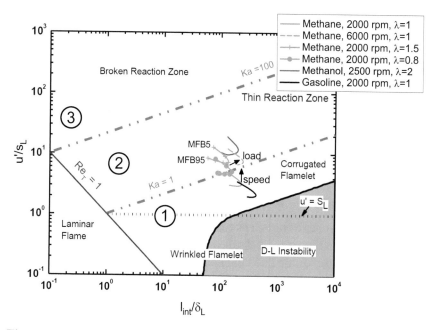

Figure 2.10: *Borghi* diagram [127] with added engine trajectories from MFB5 to MFB95, original diagram from [117], DL region from [25]. $\lambda = 0.8$ similar to a lower load at 6000 rpm

In [92], it was stated that the corrugated flamelet behavior extends beyond Ka = 1, according to [37] even up to $Ka = 40$, while [143] postulated an extension of the thin reaction zone regime to a Ka number above 100. This indicates that the *Borghi* diagram should only be used for guidance rather than giving absolute boundaries. Furthermore, internal combustion engines most likely operate in the corrugated flamelet regime. This is highlighted by the different engine trajectories shown in Figure 2.10. Even at the most critical operating point, the engine trajectory is close to the corrugated flamelet limit and below

Ka = 40. This knowledge is important when comparing different models for the turbulent flame speed for the application in engine simulation (see Chapter 4.2), as not all models are valid for all the regimes.

Figure 2.10 also highlights the conditions where the DL instability had been stated to be relevant in [25], whereas more recent literature like [150] identified a larger region of influence. With that, an influence of DL instability on the combustion in internal combustion engines could be possible and thus will be investigated in Chapter 3.1.

2.4.3 Models of the Turbulent Flame Speed

Usually, approaches to model the turbulent flame speed are either based on measurement data ([26], [106]) or theoretical considerations (e.g. *Zimont* [153]). Numerous models exist and were compared in several studies such as [92], [126] and [40]. The most comprehensive comparison was done by *Lipatnikov et al.* [92]. There, it was stated that all measurements of turbulent flame speeds show similar trends, despite a large variety of measurement methods, which also have substantial quantitative discrepancies. These discrepancies are not only caused by different (and sometimes unknown) definitions of the turbulent flame speed (see Chapter 2.4.1), but also by different flame geometries, such as spherical, Bunsen-burner or V-shaped. In consequence, turbulent flame speed models can only be evaluated on the basis of their qualitative reaction to boundary conditions changes, while their output remains quantitatively uncertain. As a result of their evaluation, *Lipatnikov et al.* reported the two models developed by *Zimont* [153] and *Peters* [117][119] to be qualitatively valid, among others. In a study published by *Burke et al.* [40], the turbulent flame speed model developed by *Muppala et al.* [109] was rated best. However, *Burke et al.* also stated that it cannot be interpreted as a general statement, since the experimental data basis of turbulent flame speeds needs to be expanded in order to ensure a reliable model evaluation. *Burke et al.* furthermore highlighted the different definitions of the turbulent flame speed in various models, which was already mentioned for turbulent flame speed measurements in Chapter 2.4.1. For example, the *Zimont* model is defined at a radius in the middle of the wrinkled flame front. This radius corresponds to the turbulent flame speed obtained from pressure trace measurements. In the context of engine simulation, this translates to $dm_E = dm_b$ in eq. 2.4, since the speed of flame propagation is directly linked to the rate of heat release. In contrast, the *Peters* model defines s_T as a tip speed. There, some unburnt mass exists behind the chosen reference radius (see Figure 2.9). With that, it is related to the entrainment approach and a characteristic burn-up time (see eq. 2.2) needs to be calculated to obtain the rate of heat release in engine simulation. The summary of s_T-models given by *Burke et al.* also illustrated that all models use at least one constant calibration parameter. While for some models, this parameter might have a phenomenological background, it still underlines the general uncertainty of modeling the turbulent flame speed.

The different results of the different model comparisons already prove that it is not possible to draw a general conclusion from model evaluations on the basis of turbulent flame speed measurements. For the application of such models in engine simulation, additional uncertainties arise: as stated by *Burke et al.*, flame speed measurements are usually performed at low pressures. Their results are then transferred to engine application, which, as already discussed for laminar flame speeds in Chapter 2.3, can cause errors. Furthermore, *Lipatnikov et al.* highlighted several differences between the combustion process in internal combustion engines and measurement devices for turbulent flame speeds: in engines, the piston movement changes the integral length scale while the turbulence level changes with time and flame radius. In measurement devices like explosion bombs, these characteristics are kept constant. Moreover, the increase in temperature and pressure due to gas compression also affect engine combustion. Therefore, the turbulent flame speed models need to be evaluated on the basis of their performance in the burn rate model environment.

2.5 Fuel Influence on Flame Wrinkling

The detailed theories of *Markstein* [98] and *Lewis* number were elaborated in numerous publications ([10][11][57][69][144]). As a basic understanding of the concepts is helpful for the following investigations, they are briefly described in this subchapter.

At a flame front deformation with a convex orientation towards the unburnt mixture, a de-focusing of heat diffusion occurs (see Figure 2.11), as the heat flux is oriented towards the unburnt. On its own, this would lower the local flame speed. The mass diffusion, however, focuses on the convex structure, as the unburnt mixture diffuses towards the flame front. Again, on its own, this effect would increase the local flame speed. The *Lewis* number Le describes the relationship between both effects and is calculated as the ratio of heat diffusivity to mass diffusivity. It is a measure for the thermo-diffusive stability of a flame. For $Le = 1$, both effects cancel each other out and the flame structure is unaffected. For a mixture with $Le < 1$, the local flame speed is increased at a convexity, since the accelerating effect of mass diffusion exceeds the decelerating heat diffusion effect. At a concave deformation, the opposite is true. This leads to an increase in the overall flame surface, as illustrated in [26]. For $Le > 1$, a decrease in flame surface is caused by the opposite trends of $Le < 1$. Consequently, the turbulent flame speed increases with decreasing *Lewis* number. For laminar flames, this effect can trigger the formation of cellular structures. As highlighted in Figure 2.3, these structures disturb the measurement of the laminar flame speed.

The mass diffusion used to calculate Le is usually related to the deficient reactant. For lean mixtures, this is the fuel. According to *Matalon* [101], an effective *Lewis* number can be calculated by using a weighted average of the *Lewis* numbers of fuel and oxidizer. The weighting depends on λ. At stoichiometry, the effective Le number is the average of Le_{fuel} and Le_{oxidizer}. Preferential diffusion of fuel and oxidizer is thereby accounted for to a certain degree. According to *Law et al.* [89], preferential diffusion has an influence on the turbulent flame speed similar to thermo-diffusive effects. For example, if the deficient reactant has a high diffusivity, it tends to accumulate at concave flame structures. This shifts the local mixture composition towards stoichiometry and, with that, increases the local flame speed.

Another approach to describe the fuel influence on flame wrinkling is the *Markstein* number. As written in [89], it incorporates the effects of nonequidiffusion via the *Lewis* number and additionally the influence of stretch. The *Markstein* number is calculated by dividing the *Markstein* length by the laminar flame thickness. The *Markstein* length describes the flame speed change with stretch rate. It can be measured, for example, from spherically expanding,

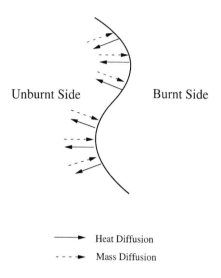

Unburnt Side   Burnt Side

——→ Heat Diffusion
- - - ► Mass Diffusion

Figure 2.11: Schematic representation of heat and mass diffusion at a wrinkled flame [89]

non-turbulent flames: when plotting the measured flame speed over measured stretch rate (proportional to the flame radius), a linear trend arises (see Figure 2.3). The slope of this trend is the *Markstein* length. The laminar flame speed is estimated by extrapolating the slope to zero stretch. In turn, the stretched flame speed s_f can be calculated by

$$s_f = s_L - \delta_M \cdot \kappa$$ eq. 2.14

where κ is the stretch rate and δ_M the *Markstein* length. According to [69], the *Markstein* length is strongly linked to the *Lewis* number. The qualitative similarity of *Markstein* numbers for different fuels from [10] and the *Lewis* number calculated using *Matalon's* approach [101], as shown in Figure 2.12 (left), highlights this strong coupling. Furthermore, it is underlined by the work of *Chaudhuri et al.* [26] and *Nguyen et al.* [106]. While *Chaudhuri et al.* succeeded in using the *Markstein* length to approximate the turbulent flame speeds of different fuels, *Nugyen et al.* obtained results of similar quality by using the *Lewis* number. As the application of the *Markstein* approach requires a (local) stretch rate, which cannot be obtained directly in the 0D/1D simulation model class, the *Lewis* approach is considered more suitable. In order to quantify the expected fuel influence on the turbulent flame speed s_T, the proportional relationship between *Lewis* number and s_T, $s_T \propto 1/Le^{0.5}$, as published in [106], is plotted in Figure 2.12 (right) for different fuels and λ. For methane, the almost constant *Lewis* number results in an almost unchanged turbulent flame speed for a variation of λ. For methanol, similar trends are observed. For ethanol and especially gasoline, however, the strong change in Le with fuel leads to a strong reduction of the turbulent flame speed with increasing Le. Furthermore, due to the λ-dependency of Le, a change in s_T with λ is observed, which is most prominent for $\lambda < 1.2$.

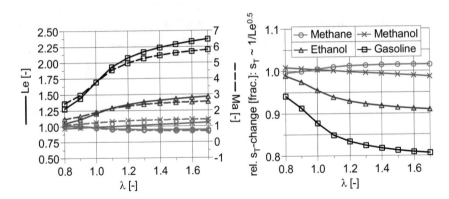

Figure 2.12: Left: comparison of *Lewis* number Le (calculation method from [101]) and *Markstein* number Ma from [10]; right: expected influence of Le effects on s_T from [106]. Source: [72]

It has to be noted that turbulent flame speeds are usually measured at a pressure of 20 bar or less and only moderate temperatures. In [89] and [51], the influence of pressure on the *Markstein* length was investigated. For different fuels and λ, the *Markstein* length tends to become zero with increasing pressure. In [119], it was stated that the flame surface increase by turbulence is the main phenomenon influencing s_T, while effects like nonequidiffusion play a minor role. This theory is supported by findings published in [112]. As turbulence levels in an engine are high compared to measurements of s_T and the pressure is significantly above 20 bar, the relevance of fuel influence on flame wrinkling in engine combustion cannot be derived reliably from fundamental research. Optical investigations in an engine, as published in [20], seem to show the change in flame speed with changing fuels as expected from nonequidiffusion effects, but with a reduced sensitivity compared to combustion bomb measurements. Furthermore, the optical engine was operated at a constant ignition point. This caused the start of combustion to change with changing fuel, probably because of the different energies needed to ignite the air-fuel mixture. Since turbulence decreases with crank angle degree, and the fuels with the highest expected Le influence ignited the latest, the difference in turbulent flame speed might be related to a change in turbulence level. In consequence of these uncertainties, the influence of changing fuels on combustion was investigated systematically using engine measurements in the course of this work. Additionally, as described in Chapter 2.3.2, the hydrodynamic instability may also affect the combustion and is therefore examined, too.

3 Measurement Data Analysis

In order to evaluate the relevance of the *Darrieus-Landau* instability and the fuel-dependent flame wrinkling for engine combustion, as described in Chapter 2.3.2 and Chapter 2.5, a measurement data analysis was carried out on two engines. Their technical data and available operating conditions are listed in Table 3.1 and Table 3.2 for engine A and B, respectively.

Table 3.1: Engine A: technical data and operating conditions

Bore [mm]	92.9
Stroke [mm]	86
CR [-]	13
Load (IMEP [bar])	11
Speed [rpm]	1500, 2000, 3000
λ [-]	0.8 to 1.7
Fuel	Methane, ethanol, gasoline

Table 3.2: Engine B: technical data and operating conditions

Cylinder no.	2
Bore [mm]	89
Stroke [mm]	68
CR [-]	10.5
Load (IMEP [bar])	0 to 22, full engine map
Speed [rpm]	2000 to 6000, full engine map
λ [-]	0.8 (gasoline) to 1
Fuel	Methane, gasoline

While the measurement data of engine A were taken from [38], Robert Bosch GmbH kindly provided the data of engine B. For both engines, models to perform a pressure trace analysis (PTA) were set up in GT-Power [52]. The FKFS UserCylinder Plug-In [64] was used to calculate the in-cylinder processes. The location of the pressure transducers on the intake and exhaust side represented

© The Author(s), under exclusive license to
Springer Fachmedien Wiesbaden GmbH, part of Springer Nature 2021
S. Hann, *A Quasi-Dimensional SI Burn Rate Model for Carbon-Neutral Fuels*,
Wissenschaftliche Reihe Fahrzeugtechnik Universität Stuttgart,

the model boundaries. For each engine operation point, several hundred single working cycles were recorded and averaged. The averaged in-cylinder pressure traces were filtered with a second-order *Butterworth* filter with a cut-off frequency of 2500 Hz. The in-cylinder pressure trace was corrected by using the cumulative net heat release criterion as a pressure pegging method. To calculate the in-cylinder mass, the 100 % iteration method was employed with λ held constant and considering the measured emissions. Its deviation was small compared to the measured mass flow rate. The wall heat transfer approach published by *Bargende* in [8] was used to calculate the heat flux from the cylinder gas to the walls. The heat transfer from the cylinder walls to the cooling liquid was modeled in GT-Power. The calorific properties were calculated with the FKFS model published in [65] and [63]. More details on performing a PTA were given by *Fandakov* in [42].

3.1 Influence of the *Darrieus-Landau* Instability

For the investigation of the influence of the DL instability on engine combustion, a load variation at 4000 rpm of engine B fueled with methane was used. The basis for this investigation was the calculation of a turbulent flame speed. By assuming $dm_b = dm_E$ in eq. 2.4, s_T can be calculated by

$$s_{T,PTA} = \frac{dm_b}{\rho_u \cdot A_{fl}}.$$

                                                                                    eq. 3.1

dm_b and ρ_u were obtained from a pressure trace analysis (PTA). The flame surface area was tabulated dependent on the cylinder volume V_{cyl} and the relative burnt volume $V_{b,rel} = V_b/V_{cyl}$ by using the detailed combustion chamber geometry. Both volumes were taken from PTA results, too. For the load variation, $s_{T,PTA}$ only changed with CAD (see Figure 3.1).

As described in Chapter 2.3.2, a disturbance of a flame front has to have a significantly larger length scale than the laminar flame thickness for the DL instability to occur. The largest turbulent length scale is the integral length scale. From 3D CFD calculations performed in [50] using engine B, an integral length scale of about 1 mm and above was derived (see Figure 4.19). This is in accordance with values published in [120]. The laminar flame thickness can

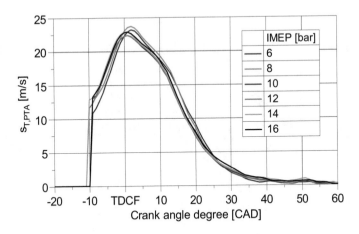

Figure 3.1: $s_{T,PTA}$ at different loads, engine B, methane, 4000 rpm [72]

be calculated from reaction kinetics. More details will be given in Chapter 4.3. The integral length scale only changes with CAD. The laminar flame thickness, however, decreases with increasing pressure and temperature and is of the size 0.1 mm and below. In theory, this difference in length scale would allow the DL instability to occur. In Figure 3.2, it is shown that for a variation of load, the in-cylinder pressure traces change strongly, while the temperature traces remain almost unchanged. Hence the laminar flame thickness is reduced significantly when increasing the engine load, thus increasing the theoretical influence of the DL instability. Thereby, $s_{T,PTA}$ should increase with load, which is not true, as highlighted in Figure 3.1. One possible explanation could be a simultaneous reduction of s_T by a reduction of s_L with increasing pressure, consequently compensating the effects of the DL instability. To validate this interpretation, the change in s_T when calculated with different s_T-models (see Figure 4.7 for citations) was investigated for varying engine loads. The most conservative result was a constant s_T with changing load. This is caused by a decrease in laminar flame thickness, counteracting the reduction of laminar flame speed with increasing pressure. Other s_T-models even showed an increase in s_T with engine load. Based on these investigations, the influence of the DL instability on combustion under engine-relevant boundary conditions seems unlikely.

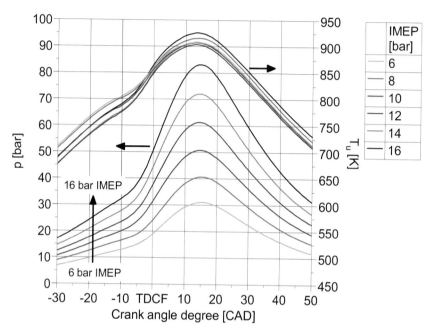

Figure 3.2: Change in pressure and unburnt temperature at different loads, engine B, methane, 4000 rpm [72] (edited)

In addition to the influence of a load variation on $s_{T,PTA}$, its general change
with crank angle degree is of interest. As depicted in Figure 3.1, a maximum
is formed about 10 CAD after combustion start. This corresponds to a mass
fraction burnt of 10 % and an approximate flame radius of 20 mm. Despite an
increase in A_{fl} and ρ_u, $s_{T,PTA}$ also increases at first. This is a result of the dispro-
portionately high increase in dm_b, which indicates an increase in flame wrinkling. Optical investigations on a geometrically similar engine, which were
performed by *Martinez et al.* [100] at an engine speed of 900 rpm, showed
a qualitative change in s_T with CAD similar to Figure 3.1: a maximum was
formed at a flame radius of 20 mm and 10 % mass fraction burnt. In literature such as e.g. [140], the increase in turbulent flame speed was related to a
growing spectrum of turbulent length scales causing a stronger wrinkling of
the flame front. However, since the size of the flame is far greater than the

largest turbulent length scale of $l_{int} = 1$ mm, the full turbulent spectrum should be effective before the maximum of $s_{T,PTA}$ is reached, given a sufficiently long time of turbulence-flame interaction [20]. Another explanation for the $s_{T,PTA}$-increase could be the increase in local turbulence at small flame radii, as plotted in Figure 4.17 and elaborated in Chapter 4.5. In their optical investigations, *Martinez et al.* [100] furthermore observed a movement of the flame centroid with CAD. *Martinez et al.* related this effect to a non-uniform flame propagation and a movement of the whole flame due to the tumble flow. In Figure 4.18, it is illustrated that the distribution of the turbulent kinetic energy does not change significantly with CAD (see Figure 4.18). With that, the flame centroid movement can also affect the local turbulence acting on the flame. Additionally, multiple single working cycles were averaged to determine s_T in both PTA and optical investigations. This adds the cycle-to-cycle variations as an influence on the shape of s_T. More details on this subject will be given in Chapter 4.4.2.

It is important to note that the optical access of the combustion chamber in [100] only covered about 60 % of the bore area. To determine the flame speed, a reference circle was calculated on the basis of the detected flame area. The growth rate of the reference circle yields the flame speed. If the limit of the optical access is exceeded, the detected flame surface does not correspond to the real flame surface anymore. The real flame is cut off at the optical access limit, consequently causing s_T to decrease. As the maximum of s_T occurred before the optical limit was reached, the following decrease of s_T in both the PTA and the optical investigation most likely results from a general decrease of TKE with CAD. Furthermore, s_T decreases even more due to the reduced turbulence in the vicinity of the cylinder wall. This effect is comparable to the cut-off of the flame at the optical access limit. In addition to validating the calculated $s_{T,PTA}$, the optical investigations give an initial indication of an influence of the local TKE on combustion, which will be elaborated in Chapter 4.5.

3.2 Influence of Fuel on Flame Wrinkling

For the investigation of a possible fuel influence on flame wrinkling, $s_{T,PTA}$ was calculated again. Here, a variation of λ for methane, ethanol and gasoline was investigated using measurement data from engine A. The burn duration served as a second indicator for the general speed of combustion. Furthermore, the qualitative influence of Le on s_T, as highlighted in Figure 2.12, needs to be kept in mind. By its influence, a shifting of the maximum s_T towards rich mixtures, beyond the maximum of s_L, was to be expected for gasoline and ethanol. Furthermore, $s_{T,gasoline}$ was expected to be significantly lower than $s_{T,ethanol}$ due to the lower laminar flame speed (see Appendix A1.1) and the much stronger influence of Le (see Figure 2.12). However, none of these expectations were confirmed by the measurement results, which are displayed in Figure 3.3. The minimum burn duration and maximum s_T was reached for the same λ as was the case for s_L (see Figure 2.7 and Appendix A1.1). Furthermore, gasoline and ethanol had very similar burn durations and similar burn duration dependencies with respect to λ. This contradicts the much stronger s_T-reduction for increasing λ of gasoline compared to ethanol, as depicted in Figure 2.12.

The late-shifting of MFB50 for gasoline is of special interest. It was necessary in order to avoid severe engine knock. A late-shifting of MFB50 generally leads to a combustion occurring at later CAD, and, with that, much lower turbulent kinetic energy (TKE). The turbulent flame speed is thereby reduced, which can be seen in the bottom left graph of Figure 3.3. In consequence, the difference between the calculated turbulent flame speed of gasoline and ethanol at similar MFB50 can be expected to remain at the same level if MFB50 would have been kept constant when varying λ. This trend would be very similar to that of s_L for different fuels at varying λ (see Appendix A1.1). Since the influence of Le on $s_{T,gasoline}$ remains constant for $\lambda > 1.2$, the expected Le-induced s_T-reduction cannot be the reason for the trend seen in Figure 3.3. Furthermore, this trend shows the high sensitivity of the turbulent flame speed with respect to the turbulent kinetic energy. Appendix A3.1 illustrates that all the trends observed at 2000 rpm are also valid for measurements at 1500 rpm and 3000 rpm. In conclusion, none of the trends of fuel influence on flame wrinkling, as expected from turbulent flame speed measurements (see Figure 2.12), could be observed in engine measurement data. This is in accordance with [119], where the majority of flame wrinkling was expected to be caused by the interaction of flame and turbulence, instead of fuel-dependent influences. A possible explanation for the lack of a fuel influence on flame wrinkling could be the difference in boundary conditions between measured s_T and engine combustion (see Chapter 2.5), which seems to influence the phenomena dominating the flame wrinkling process. In consequence, a fuel influence on flame wrinkling does not need to be considered in engine simulation. Therefore, heat release rate changes due to a fuel variation can be modeled by a variation of laminar flame speed and laminar flame thickness alone. This results in the tasks of reliably modeling the laminar flame speed and thickness as well as finding a s_T-model that properly reacts to changes in turbulence, s_L and δ_L. These tasks will be addressed in the following chapters.

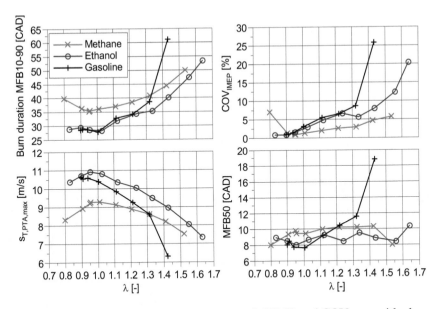

Figure 3.3: Change in burn duration, $s_{T,PTA}$, MFB50 and COV$_{IMEP}$ with changing λ and fuel, 2000 rpm, engine A [72] (edited)

4 Burn Rate Model Improvement

In its original specification, the burn rate model described in Chapter 2.2 only allows for a prediction of moderate fuel variations, like adding hydrogen to methane (see [74]). For a more significant fuel variation like methane, gasoline and ethanol, a recalibration is necessary (see Figures 5.2, 5.3 and 5.4). The reason for the needed recalibration is either the burn rate model itself or a missing, fuel-dependent influence. In fundamental research, a fuel influence on flame wrinkling is described and not yet considered in the burn rate model. However, Chapter 3 indicated that this influence seems to be small to non-existent in engine combustion. Based on these findings, the burn rate model is improved in the following.

4.1 Laminar Flame Speed Model

In conclusion of the findings presented in Chapter 3, the influence of a fuel variation on engine combustion primarily results from a change in laminar flame speed s_L and laminar flame thickness δ_L. For this reason, a reliable model is needed to cover the influence of varying boundary conditions such as temperature, pressure, fuel, air-fuel ratio, water injection and EGR on the laminar flame speed. An explanation of these effects was already given in Chapter 2.3.3. As measurements of laminar flame speeds are limited to relatively low pressures and temperatures (see Chapter 2.3.1), reaction kinetics calculations were used to obtain reference values at engine-relevant boundary conditions for the laminar flame speed model development process. Methane, CNG, ethanol, gasoline, methanol, methyl formate, hydrogen and DMC+ (65 vol−% dimethyl carbonate, 35 vol−% methyl formate) were considered as possible fuels. Pure fuels like ethanol can be modeled directly in reaction kinetics calculations. As gasoline consists of a large variety of different fuel components, whose exact fractions are usually unknown, a surrogate fuel is needed. The following identification of a suitable surrogate fuel for gasoline was performed by *Nußbaumer* in [111], under the supervision of the author.

4.1.1 Surrogate Fuel for Gasoline

[71] In general, either primary reference fuels (PRF) or toluene reference fuels (TRF) serve as surrogate fuels to model gasoline in reaction kinetics calculations. In PRF, iso-octane and n-heptane are, per definition, mixed linearly for a certain octane number. At high octane numbers, this leads to a hydrogen-to-carbon (H/C) ratio close to that of iso-octane with a value of 2.25, whereas real gasoline fuels usually have an H/C ratio of approximately 1.75 – 1.9 [9]. To match this fuel property, TRF additionally includes toluene (C_7H_8). This way, the properties of aromatic hydrocarbons are accounted for in the reference fuel. Concerning flame speed calculations, the H/C ratio is of importance, as it defines the amount of air needed to burn the fuel at stoichiometric conditions. This has a direct influence on the fuel mole concentration in the mixture. Furthermore, the ratio of CO_2 to H_2O in the exhaust gas is changed, which is of importance when investigating the influence of exhaust gas recirculation (EGR). Therefore, the use of a TRF is expected to be more accurate. Concerning engine knock, a TRF allows to model fuels of different sensitivities *S*, which describes the difference between research octane number (RON) and motor octane number (MON).

The importance of including toluene in a surrogate fuel becomes apparent when comparing calculated laminar flame speeds of iso-octane, n-heptane and toluene at an unburnt gas temperature T_u of 1000 K and a pressure *p* of 100 bar, as illustrated in Figure 4.1. Here, the reaction mechanism presented by *Cai et al.* [22] was used for calculations. At stoichiometric conditions, for example, the flame speeds of n-heptane and toluene are significantly higher compared to iso-octane. When using a PRF at a typical RON of 98, which contains 2 vol-% of n-heptane, the laminar flame speed s_L is only slightly different compared to iso-octane. A TRF with the same RON, a MON of 89 and a H/C ratio of 1.85, which represents a typical "Super"-gasoline as defined in the DIN EN 228-standard [32], is composed of 66.59 vol-% iso-octane, 8.98 vol-% n-heptane and 24.43 vol-% toluene. This composition was obtained by applying the calculation rule presented in [22]. Due to the higher amount of n-heptane and toluene in the TRF, its flame speed is higher than that of iso-octane and, with that, the PRF. This result is in accordance with measurements published in [34], where gasoline showed a significantly higher laminar flame speed than iso-octane.

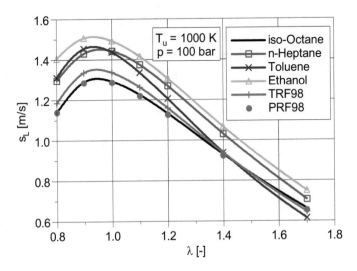

Figure 4.1: Comparison of gasoline reference fuels with pure components [72] (edited)

As stated above, the method to calculate the TRF composition is based on the values of RON, MON and H/C ratio. Therefore, a representative TRF has to be defined. For this purpose, fuels in the range of RON = 69 – 102, MON = 66-89.1 and H/C ratio = 1.743 – 2.08 were investigated. Each pairing of RON, MON and H/C ratio represented a fuel defined in the literature [96] [81] [113]. Despite the wide range of octane numbers, the maximum difference in flame speed was approximately 3 % to 5 %, which is below measurement uncertainty. In a more detailed investigation of representative fuels with typical RON values of 95, 98, 100 and 102 in a wide range of boundary conditions (T_u = 300 K to 1200 K, p = 1 bar to 100 bar, $0.6 \leq \lambda \leq 1.7$), a maximum difference of below 2 % was observed. This small influence of RON and MON on flame speed is caused by the TRF composition of the individual fuels. With increasing octane number, the amount of n-heptane decreases, which would consequently decrease the flame speed. However, fuels of high octane number tend to have lower H/C ratios, according to literature [113]. This can only be accounted for by an increasing amount of toluene, which again increases the flame speed. This leads to an overall small change in laminar flame speed. Based on these results, an octane number influence on laminar flame speed can be neglected, especially in the typical octane number range of RON = 95 to 102. As a certain influence of the H/C ratio could be observed, various TRF of similar RON and MON with H/C ratios between 1.36 and 2.06 are compared in Figure 4.2.

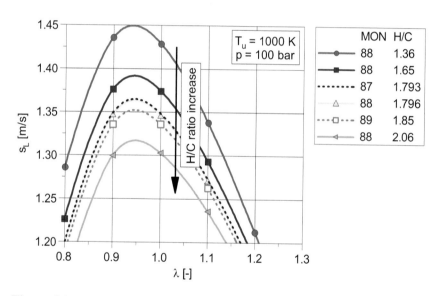

Figure 4.2: Influence of H/C ratio on s_L at RON 98 and (almost) constant MON [72] (edited). Fuels from [128][142][24][48][96]

Since the toluene amount in the TRF is the main way of influencing the H/C ratio, high toluene amounts are needed to achieve low H/C ratios. To keep the octane number constant, higher amounts of n-heptane are needed to compensate for the positive effect of toluene on octane number. Both n-heptane and toluene increase the flame speed compared to iso-octane, which effects a notable relationship between H/C ratio and laminar flame speed. However, within the general H/C ratio range of approximately 1.75 – 1.9 [9][113] at relevant octane numbers, the influence is small and can thus be neglected. Therefore, a TRF with a RON of 98, a MON of 89 and a H/C ratio of 1.85 is used as a reference fuel for laminar flame speed calculations of gasoline and is named TRF98.

Figure 4.1 also shows that ethanol has a higher flame speed than n-heptane or toluene, especially at rich conditions. Compared to the TRF98, the influence is even stronger. Since the admixture of various amounts of ethanol to gasoline is a common measure worldwide, like the E10 fuel (gasoline with 10 vol–% ethanol) in Germany, E85 in the USA or pure ethanol in Brazil, its influence on flame speed should be accounted for in engine simulation. A possible method for this will be described in Chapter 4.1.3 and evaluated in Chapter 4.1.5. However, with a RON of 109 and a MON of 90 [77], ethanol also increases the octane number of the mixture. According to [5], this advantage is exploited by using a RON 92 gasoline for the production of the E10 fuel, resulting in a RON of approximately 95, which is comparable to that of DIN EN 228 [32] "Super" gasoline. As already stated above, this small change in RON will only cause a negligible change in laminar flame speed. Hence, ethanol/gasoline-blends can be represented by mixtures of TRF98 with ethanol, without accounting for changing gasoline qualities with varying ethanol content. This way, even fuels like E85 (ASTM D5798-specification [3]), which contains ethanol of 51 % to 83 % by volume, can be modeled using the same approach.

4.1.2 Surrogate Fuel for CNG

Similar to gasoline, CNG consists of various different components. Typical compositions are given in Table 4.1. In contrast to gasoline, the fraction of these components within the fuel can be determined exactly. Consequently, no surrogate fuel is needed to model CNG in reaction kinetics calculations. With the composition of CNG being dependent on its origin (see Table 4.1), and these different CNG are mixed in the gas grid, the actual CNG composition fluctuates continuously. To evaluate the resulting change in laminar flame speed, the influence of each component is investigated individually in Figure 4.3.

Table 4.1: Composition of CNG [53] [54] [55] [56] and LNG [130]

Component [mol-%]	North Sea H	Russia H	Netherlands L	Weser Ems L	LNG Yemen
Methane	86.25	97.79	83.16	87.61	93.17
Nitrogen	0.93	0.82	10.08	9.09	0.02
Carbon dioxide	1.91	0.09	1.57	2.52	0
Ethane	8.56	0.88	4.04	0.72	5.93
Propane	1.89	0.29	0.81	0.04	0.77
n-Butane	0.39	0.1	0.23	0.02	0.12

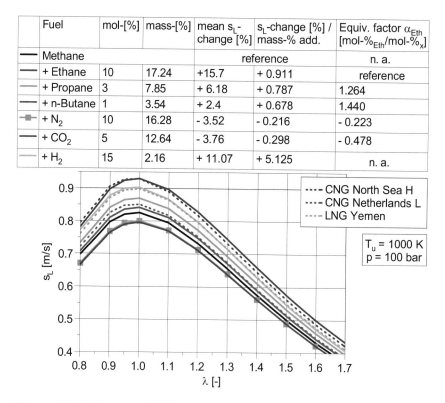

	Fuel	mol-[%]	mass-[%]	mean s_L-change [%]	s_L-change [%] / mass-% add.	Equiv. factor α_{Eth} [mol-%$_{Eth}$/mol-%$_x$]
——	Methane			reference		n. a.
——	+ Ethane	10	17.24	+15.7	+ 0.911	reference
——	+ Propane	3	7.85	+ 6.18	+ 0.787	1.264
——	+ n-Butane	1	3.54	+ 2.4	+ 0.678	1.440
-■-	+ N_2	10	16.28	- 3.52	- 0.216	- 0.223
——	+ CO_2	5	12.64	- 3.76	- 0.298	- 0.478
——	+ H_2	15	2.16	+ 11.07	+ 5.125	n. a.

Figure 4.3: Influence of individual CNG components on the laminar flame speed of methane and comparison with real fuels

Of all the components considered here, ethane has the most significant influence on the laminar flame speed. Per mass-% ethane, the laminar flame speed is increased by 0.911 %. The evaluation is performed on the basis of mass-%, as s_L increases linearly with mass-% for binary fuel mixtures (see Chapter 4.1.3). With an s_L-increase of 0.787 % per mass-% fuel, the influence of propane is second highest. In contrast to ethane, propane is usually only present in small amounts. The same applies to n-butane. Nitrogen can be present at higher amounts. Considering the high nitrogen content of the air, the additional nitrogen in the fuel only causes an insignificant reduction of s_L. Due to the higher heat capacity of CO_2 (triatomic molecule), its reducing influence on the laminar flame speed exceeds that of N_2. In CNG, the effects of the inert gases and higher hydrocarbons counteract each other to a certain degree.

Considering the results illustrated in Figure 5.5, where the change of a HRR due to a s_L-increase by 10 % is shown to be small, it is reasonable to use pure methane as a surrogate fuel for CNG or LNG. Based on the influence of each secondary component, this replacement is sufficient for a total amount of higher hydrocarbons of 6 mol–% and below, resembling a s_L-change of approximately 10 % at most. If nitrogen or carbon dioxide are present in the fuel, their amount can be subtracted from the higher hydrocarbon amount in a simple approach. The similar laminar flame speeds of methane and CNG Netherlands L support this method. At similar fractions of higher hydrocarbons, but due to the missing nitrogen and carbon dioxide content, the laminar flame speed of LNG Yemen is higher compared to CNG Netherlands L. It represents the limit of a 10 % change in s_L at a fraction of higher hydrocarbons of 6 mol–%. For even higher amounts, the ethane fraction needs to be considered as a secondary component (see also Chapter 4.1.3 and Chapter 4.1.5). A more detailed investigation can be performed by calculating an ethane-equivalent for all higher hydrocarbons and inert gases. As will be described Chapter 4.1.3, the laminar flame speed of a binary fuel mixture can be calculated by a linear interpolation between both fuels, related to mass-%. For small amounts of non-methane components in CNG, a linear relation between mol-% and mass-% can be estimated with a coefficient of determination R^2 of 0.999. This approach is valid for up to 10 mol–% ethane, 5 mol–% propane, 3 mol–% n-butane, 10 mol–% nitrogen and 5 mol–% carbon dioxide. These limits are sufficiently high for typical CNG compositions. Based on this linearity and the mean s_L-change related to mass-% of an additional component (column 3 in Table 4.1), an ethane equivalence factor α_{Eth} can be calculated (last column in Figure 4.3). For the propane fraction of CNG North Sea H (see Table 4.1) for example, this gives an ethane-equivalent of 2.39 mol-% and a total ethane-amount of 10.95 mol-%. The similar laminar flame speed of CNG North Sea H and methane with 10 mol-% of ethane, as shown in Figure 4.3, proves the validity of this method. With an ethane-equivalent of 0.913 mol-% for CO_2 and an ethane-equivalent of 0.207 57 mol-% for N_2, the overall ethane-equivalent is 9.83 mol-% and explains the slight difference between the laminar flame speed of CNG North Sea H and methane with 10 mol-% of ethane. In summary, methane is an adequate surrogate for most CNG compositions. For more detailed investigations, a mixture of methane and ethane can be used.

Figure 4.3 also investigates the influence of hydrogen admixture. Besides renewable methane, electrolytically produced hydrogen can help to reduce the overall CO_2 emissions when using CNG as fuel. When adding 15 mol-% of hydrogen, the laminar flame speed is increased by about 11 %. According to [108], the gas grid in Germany, for example, is able to hold 5 mol-% of hydrogen. This limit is planned to be expanded up to 10 mol-% of hydrogen. Within this range, the change in s_L with hydrogen is small. However, the effect of hydrogen admixture to methane will be considered in the following chapters to allow for a detailed investigation of this measure to reduce CO_2 emissions.

4.1.3 Binary Fuel Mixtures

Chapter 4.1.1 and Chapter 4.1.2 highlighted the necessity to not only model the laminar flame speed of pure fuels, but also of mixtures of gasoline and ethanol as well as mixtures of methane and ethane or hydrogen. In [132], different mixing rules for the laminar flame speed were compared, including a linear interpolation related to fuel mass fraction, originally proposed in [138]. Figure 4.4 exemplarily proves this relationship at $T_u = 800\,K$ and $p = 50\,bar$ for the fuel mixtures of interest in this thesis. In contrast, the s_L-change with mol-% is strongly non-linear. While the laminar flame speed of the pure fuels can be used as lower and upper s_L-limit for mixtures of gasoline and ethanol as well as methane and ethane, the upper limit for a mixture of methane and hydrogen has to be set at 40 mol-% H_2. Below this limit, the increase in s_L

related to mass-% is linear, while the increase related to mol-% is non-linear, as highlighted in the graph detail in Figure 4.4. Above this limit, the increase in s_L is non-linear, even related to mass-%. According to [33], this results from a shift of methane-dominated to a hydrogen-dominated combustion chemistry. In consequence, a model of the laminar flame speed of methane mixed with 40 mol-% hydrogen is needed, in addition to a model of the pure fuels.

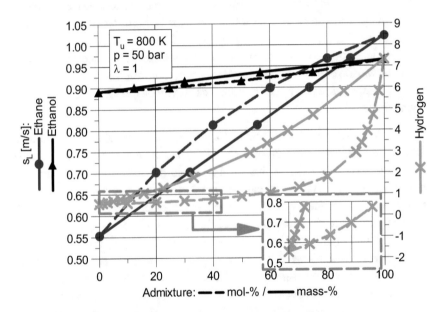

Figure 4.4: s_L-increase due to the admixture of ethane and hydrogen to methane and ethanol to gasoline, related to mol-% and mass-%

4.1.4 Model Development

In the first step of developing a laminar flame speed model, the widely used models of *Gülder* [68] and *Heywood* [75] were compared with reaction kinetics calculations. This comparison allowed to identify several deficits of both models. As highlighted in Appendix A4.1, both *Heywood* and *Gülder* showed a divergent modeling of the influence of temperature and pressure on

s_L. While *Gülder* only slightly underestimates the influence of λ, *Heywood* heavily overestimates this influence and even calculates negative s_L-values of s_L at $\lambda > 2.15$. The value of λ at maximum laminar flame speed diverges from reaction kinetics calculations in both models. Furthermore, from eq. 2.7, it can be concluded that both models calculate negative laminar flame speeds for an EGR rate above 40 %, independent on temperature or pressure. This behavior contradicts the trends obtained from reaction kinetics calculations and displayed in Figure 2.8. Another deficit was highlighted by *Konnov* in [87]: the exponents α and β in eq. 2.7 should be dependent on $\Phi = 1/\lambda$ in a non-monotonic way. While *Heywood* at least shows a monotonic dependency, *Gülder* does not consider any dependency. Furthermore, *Konnov* stated that the temperature and pressure influence should be calculated dependent on each other.

Due to these deficits of *Gülder* and *Heywood*, a different flame speed model originally developed by *Müller et al.* [107] and refined for a wider λ-range by *Ewald* [41] was used in this thesis. Since this model is based on reaction kinetics considerations, it can be regarded as semi-phenomenological and promises a higher quality fit of reaction kinetics calculations. An example for the semi-phenomenological characteristic is the limited range of validity of eq. 4.1, which is only defined for $T_b > T_0$, where T_0 is the temperature of the reaction zone (see Figure 2.2). This way, the model considers the existence of a flammability limit (see Chapter 2.3.4). However, T_0 does not correspond to the actual reaction zone temperature and serves as a calibration parameter, which is in accordance with [107]. The model calculates the laminar flame speed with following equations.

$$s_L = A(T_0) \cdot Y_{\text{react}}^m \cdot \left(\frac{T_u}{T_0} \right)^r \cdot \left(\frac{T_b - T_0}{T_b - T_u} \right)^n \qquad \text{eq. 4.1}$$

$$A(T_0) = F \cdot e^{-\frac{G}{T_0}} \qquad \text{eq. 4.2}$$

$$Y_{\text{react}}(Z^*, Y_{\text{EGR}}) = \left(\frac{Z^*}{1.1 \cdot Z^*} \right)^{n_a} \cdot \left(\frac{1 - Z^*}{1 - 1.1 \cdot Z^*} \right)^{\left(\frac{1}{1.1 \cdot Z_{\text{st}}^*} - 1 \right) \cdot n_a} \qquad \text{eq. 4.3}$$

$$\cdot (1 - Y_{\text{EGR}})^{\frac{n_{\text{EGR}} \cdot d_{\text{yH2O}}}{d_{\text{yH2O}} + y_{\text{H2O}}}}$$

$$Z^* = \frac{m_{\text{fuel}}}{m_{\text{fuel}} + m_{\text{air}}} = \frac{1}{1 + \lambda \cdot L_{\min}} \qquad \text{eq. 4.4}$$

$$T_0 = T_u \cdot S_1(Z^*) + \frac{E_i \cdot S_2(Z^*)}{\ln\left(\frac{B_i}{p}\right)} \qquad \text{eq. 4.5}$$

$$T_b = T_u \cdot (S_4(Z^*) \cdot (1 - Y_{\text{EGR}}) + Y_{\text{EGR}}) + (1 - Y_{\text{EGR}})^c \cdot S_3(Z^*) \qquad \text{eq. 4.6}$$

$$s_{\text{L,H2O}} = s_L \cdot e^{\left(-b_{\text{yH2O}} + \frac{T_u}{c_{\text{yH2O}}}\right) \cdot Y_{\text{H2O}}} \qquad \text{eq. 4.7}$$

Exponent c was added in eq. 4.6 by the author in [70][74] to better match the influence of EGR on the burnt temperature T_b, which was derived from reaction kinetics calculations. As written in Chapter 2.3.3, the injection of water not only reduces T_u, and, with that, s_L, but also causes a direct decrease of s_L by effects of dilution and heat capacity in the flame front, while chemical effects are of minor importance. The first effect is accounted for by a water evaporation model in the simulation code. The second effect is modeled by eq. 4.7 and the expanded exponent n_{EGR} in eq. 4.3. These expansions were introduced by *Crönert* in [27], under the supervision of the author. With that, the laminar flame speed model uses $T_u, p, \lambda, Y_{\text{EGR}}$ and Y_{H2O} as inputs. All other parameters not defined by any sub-equations are calibration parameters.

In [70], [74] and [71], the author already published calibrated versions of the model to match the laminar flame speed of methane, CNG, gasoline and ethanol. These versions are refined in the following and expanded to additionally cover methanol, hydrogen, methyl formate and DMC+ (65 vol–% dimethyl carbonate, 35 vol–% methyl formate). Furthermore, the method to calculate the laminar flame speed of CNG or gasoline/ethanol mixtures is changed to a linear interpolation between the pure fuels, in contrast to using additional equation for this purpose, which were published in [74] and [71]. The linear interpolation not only reduces the model calibration parameters, but also the effort of model maintenance.

Chapter 2.3.3 and Figure 2.7 already highlighted the dependence of reaction kinetics calculations of laminar flame speeds on the chosen reaction mechanism. To ensure consistency, the reaction mechanism published by *Cai et al.* [22] was employed for most fuels. Besides its wide range of validation given in the cited literature, it also shows a plausible s_L compared to additional measurements from [34], as elaborated in [111]. Although the mechanism was not developed to cover fuels like hydrogen or methanol, its results are very similar to mechanisms dedicated to those specific fuels, such as [82] and [86] for H_2, [102] and [21] for methanol, [13] and [103] for methane as well as [47] and [105] for ethanol. For methyl formate and DMC+, a different mechanism, also developed by *Cai et al.* [23], had to be used, since the *Cai et al.* mechanism [22] does not contain the species methyl formate or dimethyl carbonat.

It has to be stated that other reaction mechanisms might be just as valid as the *Cai et al.* mechanism. However, as illustrated in Figure 2.7, its calculation results represent an average laminar flame speed, making it a good starting point for the laminar flame speed model development. In Chapter 2.3.2, it was stated that reaction kinetics calculations are the only way to obtain values of the laminar flame speed at engine relevant boundary conditions. Hence, the most plausible reaction mechanism needs to be determined by evaluating the prediction quality of the burn rate model when using different mechanisms. For example, the mechanism of *Liu et al.* calculates a shifted peak value of s_L for gasoline, which could influence the change of heat release rate with λ and can be validated by comparing simulation results with engine measurements. This trend is also dependent on the influence of s_L on s_T. Therefore, the evaluation of the most plausible reaction mechanism and, with that, an implicit determination of the laminar flame speed using engine measurements is possible, but linked to the burn rate model.

4.1.5 Model Quality

The fuel-dependent calibration parameters of the laminar flame speed model for methane, gasoline, ethanol, methanol, methyl formate, DMC+, ethane, hydrogen and methane + 40 mol−% H_2 are listed in Appendices A5.1 to A5.7. Those parameters were calibrated by minimizing the least square error in combination with the standard deviation of the error between model results and reaction kinetics calculations. For these calculations, the reaction mechanism from [23] was used for methyl formate and DMC+, while the *Cai et al.* mechanism [22] was employed for all other fuels. As a starting point for the development of the latest set of calibration parameters, the data sets given by

Crönert in [27] for methane, gasoline as well as ethanol and in [28] for methanol, methyl formate as well as hydrogen were used, refined and expanded for DMC+, ethane and methane + 40 mol−% H_2. His work was performed under the supervision of the author. These data sets were the result of a continuous optimization process, starting with the parameters given by *Ewald* in [41] and their expansion to cover CNG, published by the author in [74]. *Nußbaumer* continued the model development under the supervision of the author. His results were published in [71] and [111].

Using the final set of calibration parameters, the model error of s_L is usually below 10 % at engine-relevant boundary conditions. This is highlighted in Figure 4.5. The engine-relevant boundary conditions are exemplarily identified by the temperature and pressure traces of two methane-fueled engines at different operating conditions. Appendices A6.1 to A6.3 give a more detailed overview of the error at different boundary conditions. The appendices also include the flammability limit, assumed to be at a laminar flame speed of 10 cm/s (compare Chapter 2.3.4). Laminar flame speed errors above 10 % mostly occur beyond this limit. Although the flame speed model allows to account for the flammability limit by setting $T_b = T_0$ in eq. 4.1, it is reasonable to calibrate the flame speed model beyond this limit, since its exact extent is unknown. This is supported by the laminar flame speed values at the CCV limit of different engines, as obtained in the burn rate model validation process described in Chapter 5: at the CCV limit of engine C, the laminar flame speed was about 10 cm/s to 25 cm/s for methanol or gasoline at lean air-fuel mixtures as well as for gasoline at high EGR rates. However, for engine F, the CCV limit was exceeded at flame speed values of about 5 cm/s to 15 cm/s. In contrast to engine C, this is close to the flammability limit and highlights the importance of extending the laminar flame speed model as far as possible. On a side note, the difference in s_L at the CCV limit also highlights the different effects causing high CCV, which were given in [147]: engine F was equipped with a prechamber spark plug. This should reduce the influence of local flow field characteristics and, with that, increase the relevance of flame destabilization by low values of s_L. In contrast, engine C used a regular spark plug and direct injection. Therefore, not only the local flow field at the spark plug, but also variations in the injection can increase CCV, consequently causing the CCV limit to be reached even before the values of s_L are close to the flammability limit.

In summary, Figure 4.5 and Appendices A6.1 to A6.3 prove the high model quality for different fuels, which is the prerequisite for the proper prediction of the fuel influence in engine simulation.

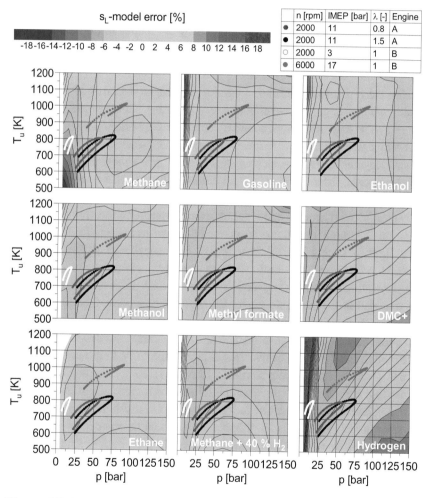

Figure 4.5: Laminar flame speed model error at $\lambda = 1$ [72] (updated and expanded)

Figure 4.6 illustrates the laminar flame speed model error for mixtures of pure fuels. E70 is a mixture of gasoline with 70 vol−% of ethanol. The laminar flame speed of the mixtures was calculated by performing a linear, mass fraction related interpolation between the modeled laminar flame speed of gasoline and ethanol, methane and ethane as well as methane and methane + 40 mol−% hydrogen, respectively. As reference, the laminar flame speed of the mixtures was determined using reaction kinetics calculations. The low model error for all mixtures proves the applicability of the interpolation method described in Chapter 4.1.3.

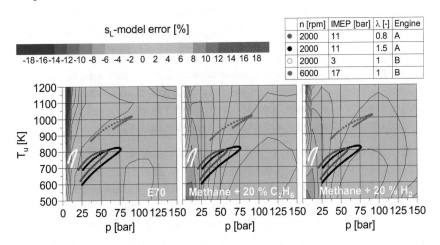

Figure 4.6: Laminar flame speed model error of fuel mixtures at $\lambda = 1$

4.2 Turbulent Flame Speed Model

To ensure a proper reaction of the turbulent flame speed model to changing boundary conditions, different models were evaluated on the basis of fundamental relationships between the turbulent flame speed and turbulence, laminar flame speed and pressure, which were described by *Lipatnikov et al.* in [92]. There, a preselection of promising s_T-models was made (see Chapter 2.4.3). In this preselection, *Lipatnikov et al.* considered the validity of the models in different regimes of the *Borghi* diagram. The *Borghi* diagram divides the characteristics of a premixed flame into three regimes: flamelet regime, thin reaction zone regime and broken reaction zone regime. More details were given in [127] and Chapter 2.4.2. In the latter, it was elaborated that according to [37] and [143], the boundaries of the *Borghi* regimes are not well known, as a flamelet-type combustion was observed at a much higher turbulence level well within the thin reaction zone regime, and the thin reaction zone regime extended into the broken reaction zone regime. In consequence, the *Borghi* diagram is more of a guide and internal combustion engines most likely operate in the flamelet regime, even for high turbulence concepts. With that, all models investigated in [92] should be applicable in engine simulation.

In order to evaluate the turbulent flame speed approach by *Damköhler* ($s_T = s_L + u'$), which is employed in the baseline burn rate model (see Chapter 2.2), its response to the fundamental boundary condition influences described in [92] is compared to more sophisticated s_T-models in Figure 4.7. The fundamental trends are used as captions of the sub-figures.

Except for the increase in s_T with u', the *Damköhler* approach reacts significantly different to boundary condition changes. Especially the very slight or even non-existing increase in s_T and ds_T/du' with s_L needs to be highlighted. Furthermore, the influence of pressure on the different s_T-models is interesting. While all models (except *Damköhler*) show the expected increase in s_T with pressure, the model developed by *Muppala et al.* [109] is the only one with an explicit pressure term in the equation. In all other models, the increase in s_T with p can be reproduced by a proper scaling of the relationship between laminar flame speed decrease and laminar flame thickness decrease. While a decrease in s_L causes s_T to decrease, a reduction of flame thickness increases the turbulent flame speed. In this context, an explicit pressure term in the equation to calculate s_T seems to be unnecessary. Hence, the *Damköhler* approach was replaced by the *Peters* approach [119][117] as a first step towards a more reliable s_T-model in engine simulation. This choice is supported by the almost constant $s_{T,PTA}$ at a variation of engine load, as depicted in Figure 3.1. While s_T increased with other models, the *Peters* approach covered the engine load influence reasonably well, indicating a plausible relationship of the s_L- and δ_L-influence in the model (see Chapter 3.1). The choice of the *Peters* model will be further justified in Chapter 4.4 and Chapter 5.

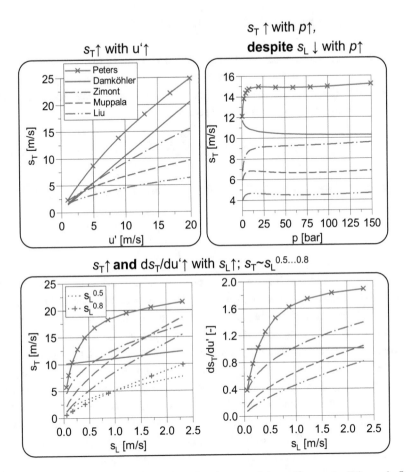

Figure 4.7: s_T-model comparison: fundamental boundary conditions influences [72] (edited). Model references: *Damköhler* [29], *Zimont* [153], *Liu* [93], *Muppala* [109], *Peters* [119][117]

4.3 Laminar Flame Thickness Model

All models of the turbulent flame speed compared in Figure 4.7 (except *Damköhler*) need the laminar flame thickness δ_L as an input parameter. However, various definitions of the laminar flame thickness exist. They are either based on flame temperature gradient [121] (eq. 4.8)

$$\delta_{L,dT,max} = \frac{T_{ad} - T_u}{\frac{dT}{dx}\big|_{max}} \qquad \text{eq. 4.8}$$

or the inner layer temperature T_0 [59] of the flame (eq. 4.9), see Figure2.2.

$$\delta_{L,T_0} = \frac{\alpha}{s_L} = \frac{\frac{\lambda}{c_p}\big|_{T_0}}{(\rho \cdot s_L)\big|_{T_u}} \qquad \text{eq. 4.9}$$

For the latter, the location of the inner layer is defined as the maximum of the H_2+CO concentration [124] or only the H_2 concentration [118]. At the respective location, the inner layer temperature T_0 and consequently the heat conductivity λ and the heat capacity c_p are evaluated. According to [66], the temperature gradient definition and the inner layer temperature definition are related, which was also observed in this study when all definitions were applied to reaction kinetics calculations. Despite a small change in boundary condition influences, the main differences between the definitions can be compensated with a linear scaling factor. Since the turbulent flame speed model published by *Peters* [117][119] was used, the flame thickness definition also published by *Peters* in [118] was chosen, where the maximum H_2 concentration defines the inner layer location.

For hydrogen, this maximum occurs in the unburnt mixture, and not in the inner layer of the flame. Thus, the flame thickness definition cannot be applied. In that specific case, the temperature to calculate eq. 4.9 was taken at the maximum heat release rate instead. For comparison, this was also done for methane. There, an almost linear relationship between both flame thickness calculations was observed. To ensure consistent definitions for the flame thicknesses of all the different fuels, the linear relationship found for methane was then applied to scale the flame thickness calculated on the basis of the maximum heat release rate for hydrogen.

The laminar flame thickness was modeled on the basis of the definition published in [119], see eq. 4.10. The fuel-dependent parameter a_{δ_L} (see Table 4.2) was added to match reaction kinetics calculation results, linearly scaled in the case of hydrogen. The kinematic viscosity v was calculated by using eq. 4.11. The dynamic viscosity η was fitted to the averaged dynamic viscosity of methane, ethanol and gasoline at different pressures, λ and EGR rates, as no significant influence of those boundary conditions was observed in reaction kinetics calculations.

$$\delta_L = a_{\delta_L} \cdot \frac{v}{s_L} \qquad \text{eq. 4.10}$$

$$v = \frac{\eta}{\rho_u} = \frac{3.85 \cdot 10^{-7} \cdot T_u^{0.6774}}{\rho_u} \qquad \text{eq. 4.11}$$

Table 4.2: Fuel-dependent values of a_{δ_L}

Fuel	a_{δ_L} [-]
Methane	1.7665
Ethanol	1.6844
Gasoline (TRF)	1.6661
Methanol	1.72684
Methyl formate	1.654
DMC+	1.652
Ethane	1.719
Methane + 40 mol−% H_2	1.95
Hydrogen	2.2

Figure 4.8 shows the difference between modeled and calculated laminar flame thickness for all fuels at stoichiometric conditions without EGR or water injection. To model δ_L, s_L from reaction kinetics calculations was used. The maximum error for most fuels is below 6 % and does not change significantly at off-stoichiometric conditions or higher rates of EGR or water injection. Since the modeling approach for the laminar flame thickness is based on the laminar flame speed model, it incorporates its error. In this context, the very low error

of the δ_L-model is advantageous. In order to investigate the influence of the s_L- and δ_L-model error, the results from reaction kinetics calculations were imported in the burn rate model instead of using the models. The resulting change in prediction quality is evaluated in Chapter 5.

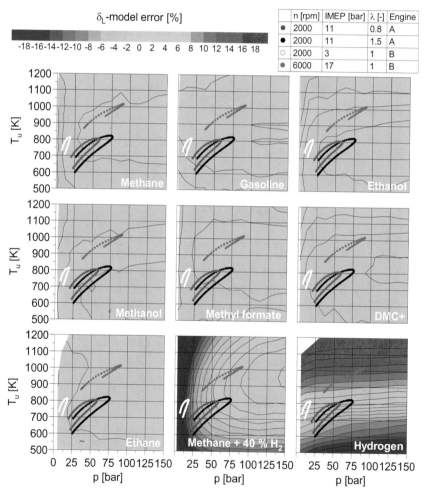

Figure 4.8: δ_L-model error of pure fuels in %, $\lambda = 1$, $Y_{EGR} = 0\,\%$, $Y_{H2O} = 0\,\%$ [72] (updated and expanded)

The error for hydrogen, also displayed in Figure 4.8, needs to be discussed separately. As stated above, the calculation method used for the other fuels cannot be applied to hydrogen. Therefore, a linear scaling was applied to a different flame thickness definition. However, the different definitions show slightly different trends for varying boundary conditions. This could be the reason for the temperature influence on the model error for hydrogen. Although the error is below ±12 % within the engine-relevant range highlighted by the dotted lines, the laminar flame thickness model for hydrogen should be re-evaluated in future work, especially when engine measurements are available.

Figure 4.9 plots the laminar flame thickness model error for different fuel mixtures. As for the laminar flame speed of fuel mixtures, a linear, mass fraction-related interpolation method was applied to calculated a_{δ_L} of the mixtures. Again, the laminar flame speed taken from reaction kinetics calculations was used as input for the δ_L-model. The δ_L-model error for fuel mixtures is very similar to that of the pure fuels. This justifies the application of the mass fraction-related interpolation method to the modeling of δ_L.

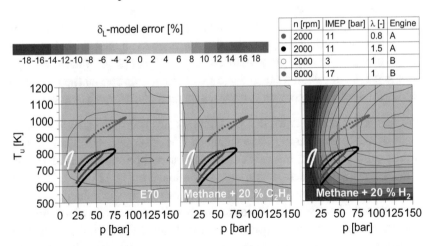

Figure 4.9: δ_L-model error of fuel mixtures in %, $\lambda = 1$, $Y_{EGR} = 0\%$, $Y_{H2O} = 0\%$

4.4 Calculation of the Characteristic Burn-up Time

On the basis of the improved sub-models for s_L, δ_L and s_T, the performance of the burn rate model was evaluated. For this, a λ-variation of methane at 2000 rpm was used. The resulting IMEP error, plotted in Figure 4.10, had a strong dependency on λ. Although the measurement data analysis, described in Chapter 3.2, indicated that a fuel influence on flame wrinkling is unlikely, methane was used here due to its unity *Lewis* number. Hence, an influence of fuel on flame wrinkling can be excluded as reason for the IMEP error with certainty. This allows the isolated investigation of the burn rate model performance.

The burn rate model can be calibrated by changing the flame surface A_{fl} (see eq. 2.4), χ_{Taylor} (eq. 2.3) and the turbulence level u'. Here, the general level and trend of the turbulence was calculated using the model from [15], which was already described in Chapter 2.2. With the turbulence level changing only very slightly vor a variation of λ, the prediction of the λ-influence on burn rate cannot be improved directly by adapting u'. However, some turbulent flame speed models show a change in the relative influence of e.g. flame thickness on s_T at different levels of u'. Since the turbulence model gives levels and trends of u' similar to CFD calculations within a small range of uncertainty, its results can be considered reliable and the indirect influence of the u'-level on the λ-trend can be an indicator for the evaluation of different s_T-models.

In Chapter 2.2, it was stated that the flame surface is usually calculated on the basis of a cylindrical combustion chamber. The spark plug position is used as a calibration parameter instead of matching the real position. Considering the flame centroid movement with time, as mentioned in [1], [2], [99], [100] and [67], this calibration method seems valid. To investigate the influence of flame surface on the heat release rate shape, the simplified approach of a cylindrical combustion chamber is compared with the detailed flame surface calculated on the basis of 3D data for engine B in Appendix A7.1. The 3D data-based calculation method was developed by *Schmid* in [131] under the supervision of the author. Due to the centroid movement, the spark plug position was varied in the 3D data, too. The resulting flame surfaces were averaged to represent an averaged working cycle. In Appendix A7.1, the heat release rate shape of the calibrated, cylindrical combustion chamber is identified as being similar to the detailed geometry. This justifies the simplification of the combustion chamber geometry. In contrast, using a cylindrical combustion chamber geometry in combination with the real spark plug position causes a rapid decrease of heat release, resulting from the simultaneous contact between flame and cylinder wall in all directions of flame propagation. This subject was elaborated in more detail in [60].

With that, only χ_{Taylor} is left to be calibrated. Target of this calibration was to match the heat release rate at $\lambda = 1$. Subsequently, a variation of λ was performed without a recalibration of the model. The IMEP error of this variation for $\chi_{\text{Taylor}} = 15$ is depicted in Figure 4.10, where a strong increase in IMEP error with changing λ can be observed.

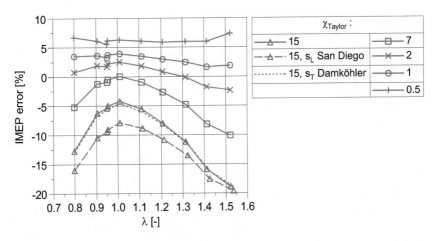

Figure 4.10: Burn rate model test with improved s_L, δ_L and s_T, engine A, methane, 2000 rpm [72]

The λ-dependent IMEP error could have multiple reasons. The change in s_L with λ could be wrong. However, Figure 2.7 illustrates that especially at lean conditions, the change in s_L is very similar for different reaction mechanisms. Nevertheless, this possible effect was investigated by using s_L calculated with the *San Diego* mechanism [102], which showed the lowest change in s_L with λ (see Figure 2.7). Without recalibrating the model, the IMEP error increased in general due to the lower s_L-value, but the trend with λ only changed slightly

(Figure 4.10, dashed line). The IMEP error might also be a result of an excessive influence of s_L on s_T. This possibility was excluded by using the s_T-model of *Damköhler*: although this model features the lowest influence of s_L on s_T (see Figure 4.7), almost no change in IMEP error was observed (Figure 4.10, dotted line). Another reason might be the calculation of u', l_{int} or A_{fl}. As u' and l_{int} only change with CAD for a variation of λ and the error does not change with different s_T-models, which have a different u'-dependent λ-influence, this cannot explain the error trend. Concerning the flame surface, the slower combustion at lean λ could theoretically change the geometric conditions during combustion. However, a variation of spark plug position or the use of detailed combustion chamber geometry did not reduce the error.

This leads to the assumption that the reason for the error is the calculation of the characteristic burn-up time τ_L (eq. 2.2). As all inputs of eq. 2.2 were excluded as a reason, the equation itself must be the issue. The very low change in IMEP error with changing s_T-models (Figure 4.10, dotted line) supports this theory due to the very strong difference in s_L-influence on s_T between both models. To investigate this assumption, χ_{Taylor} was reduced step by step, as plotted in Figure 4.10. The χ_{Taylor}-reduction decreases the influence of the characteristic burn-up time on the burn rate model. With decreasing χ_{Taylor}, the IMEP error trend with changing λ was reduced significantly, which proves the theory: with changing λ, the *Taylor* microscale changes only due to changes in the kinematic viscosity, since $v = \eta/\rho$, where v only depends on the temperature and ρ changes with λ due to the increased pressure needed to keep the IMEP constant at constant charge air temperature. As these physical relationships are fixed, the only reason for the strong λ-influence is the excessive influence of the laminar flame speed on the calculation of the characteristic burn-up time. This also explains the very strong IMEP error change when using different reaction mechanisms to calculate s_L, despite a moderate influence of s_L on s_T. From this investigation, it can be concluded that the calculation of a characteristic burn-up time τ_L is not necessary to properly predict the λ-influence on heat release rate (HRR) changes, which is shown in Figure 4.11. There, the HRR are calculated as $dQ_b/dt = \rho_u \cdot A_{fl} \cdot s_T \cdot H_{u,mix,grav.}$, instead of using the τ_L-approach. At the leanest λ, the high COV_{IMEP} of 6 % (see Figure 3.3) might cause the stronger HRR change in PTA and explain the displayed difference.

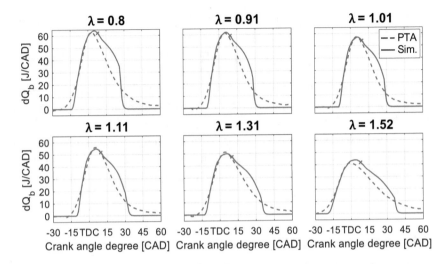

Figure 4.11: Heat release rate change without τ_L-approach, engine A, methane, 2000 rpm [72]

The same approach also allows the different s_T-models compared in Figure 4.7 to be validated. When e.g. using the *Zimont* model, the influence of λ and fuel on HRR is overestimated. In this comparison, the *Peters* model showed the best predictive behavior and was therefore used in the burn rate model proposed in this thesis. In Chapter 2.4.1, it was stated that the value of the turbulent flame speed depends on the reference radius chosen for measurements. This also applies to s_T-models, which are developed for a certain radius definition. Here, the *Peters* model is related to the tip speed definition of the turbulent flame speed, which is referenced to the widest expansion of the wrinkled flame front. With unburnt mass existing at the burnt side of this reference (see Figure 2.9), the tip speed definition is in accordance with the entrainment approach of using a characteristic burn-up time.

Without using the τ_L-approach, the influence of λ on HRR can be covered. However, as visible in Figure 4.11, the shape of the heat release rate strongly contradicts the HRR from pressure trace analysis. This highlights the reason why the τ_L-approach (eq. 2.2) is used: it has a smoothing effect on the dm_b-calculation (see eq. 2.1). As the change in boundary conditions can be predicted well without the τ_L-approach, it is reasonable to calculate τ_L by using eq. 4.12.

$$\tau_L = \frac{l_{int}}{s_T} \qquad \text{eq. 4.12}$$

This way, an overestimation of s_L in the τ_L-approach is prevented. By using the integral length scale l_{int} instead of the *Taylor* microscale l_T, no additional boundary condition influence besides s_T is included in the equation, as l_{int} only changes with CAD. Since $s_T \propto u'^x$ with $x \approx [0.3; 0.8]$ (see Figure 4.7 and eq. 4.21) and thus $\tau_L \propto 1/u'^x$, the influence of turbulence on τ_L is similar to the baseline model, where $l_T \propto 1/u'^{0.5}$ and thus $\tau_L \propto 1/u'^{0.5}$. The eq. 4.12 can be interpreted in different ways.

4.4.1 Phenomenological Interpretation

Phenomenologically, the baseline approach to calculate τ_L corresponds to a flame pocket of size l_T which is consumed in a time period dependent on s_L (see Figure 4.12). This was explained in more detail in [140]. According to its definition, the *Taylor* microscale is always smaller than the integral length scale l_{int} and larger than the *Kolmogorov* scale [122]. The integral length scale is defined as the size of the largest vortex in the flow field. In contrast, the *Kolmogorov* scale l_K [84][85] defines the smallest vortex size. These two scales define the range of vortices that wrinkle the flame front, as long as the laminar flame thickness δ_L is smaller than l_K [127]. This condition is met in the corrugated flamelet regime of the *Borghi* diagram. In [37], it was stated that the flamelet regime extends to high values of turbulence. In consequence, internal combustion engines most likely operate in the corrugated flamelet regime alone. Furthermore, the shift with increasing λ towards the thin reaction zone regime in the *Borghi* diagram, caused by an increase in laminar flame thickness and a decrease in laminar flame speed, is probably not strong enough to leave the corrugated flamelet regime. When looking at the schematic representation of the integral length scale l_{int}, the *Kolmogorov* scale l_K and the *Taylor* microscale l_T in Figure 4.12, it is apparent that flame wrinkling is still present at length scales below l_T. Since the baseline model assumes a flame speed of s_L at vortex scales of l_T, it neglects flame wrinkling at length scales below l_T. As was shown above, this leads to an overestimation of the influence of s_L in the burn rate model, which is especially relevant for strong changes of λ or fuel, for example. In contrast, the improved model (Figure 4.12) calculates the characteristic burn-up time according to eq. 4.12. The full scale of flame wrinkling is thereby accounted for in the model approach (see Figure 4.12) and the overestimation of the s_L-influence on dm_b is corrected. In this context, the definition of the turbulent flame speed is important. The phenomenological concept of flame pockets being consumed requires a turbulent flame speed that is defined as the tip speed of the flame front. The turbulent flame speed model of *Peters* used in this study fulfills this condition.

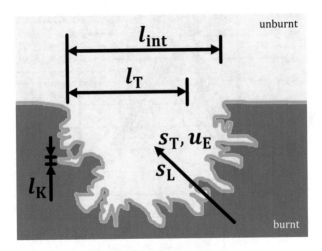

Figure 4.12: Phenomenological model comparison [72]. s_L, l_T: baseline model [133]; s_T, l_{int}: improved model

4.4.2 Mathematical Interpretation

Another way of interpreting the τ_L-approach is a mathematical one, also presented in [140]. There, it was stated that the differential equations of the τ_L-approach are needed to achieve the typical shape of mass fraction burnt by creating a delay between dm_E and dm_b. This interpretation is continued in the following. If the burn rate is calculated without using the τ_L-approach ($dm_E = dm_b$ in eq. 2.4), a rugged burn rate shape is observed (see Figure 4.11). This underlines the necessity of the τ_L-approach. When looking at heat release rates of single working cycles (SWC) from PTA (see Figure 4.13), their rugged shape is similar to those seen in Figure 4.11. Furthermore, the rugged shape is changed by the influence of cycle-to-cycle variations (CCV). Despite the strong difference between the HRR of different SWC, the average HRR of only 30 SWC is very close to that of 500 SWC (see Figure 4.13). The great similarity between HRR of SWC and simulated HRR without τ_L-approach leads to the interpretation that the τ_L-approach is needed to model the HRR of an averaged working cycle (AWC) by smoothing the HRR of a SWC.

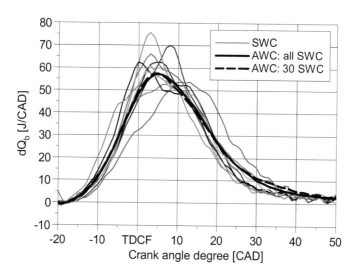

Figure 4.13: Heat release rate of single and averaged working cycles from pressure trace analysis [72]

Figure 4.14 shows several HRR of SWC generated in simulation by varying the ignition point and the laminar flame speed. If these HRR are averaged using a weighting function, an artificial HRR of an AWC, represented by thick broken line, is obtained. A normal probability density function (PDF) was used for weighting the SWC. The mean value and standard deviation of the PDF were taken from the MFB50 distribution of the SWC measurements. Compared to the AWC from PTA (thick solid line in Figure 4.14), an overall great similarity is observed, despite the simple weighting approach. A far greater similarity is expected to be achieved when using a more sophisticated averaging method and simulating the SWC more accurately. Nevertheless, the good result when using the simple approach already proves that the shape of an averaged working cycle can be obtained by simulating SWC and accounting for CCV influences. This underlines the interpretation of the τ_L-approach as method to generate a HRR of an AWC by using one SWC. In this context, it is again reasonable to calculate τ_L on the basis of l_{int} and s_T in order to avoid the addition of cross-influences due to the necessity for τ_L.

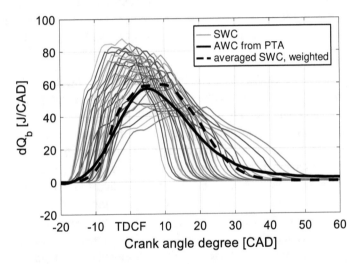

Figure 4.14: Heat release rate of single and averaged working cycles from simulation compared to PTA results [72]

Furthermore, it can be concluded that a simulated HRR when using the τ_L-approach corresponds to a certain CCV level. If the CCV level changes, the comparability of measurement and simulation is limited. This theory will be supported by the validation using engine A (high CCV change) and engine C (almost no CCV change), described in Chapter 5. Consequently, a model is required to take CCV influences like the reduction of engine efficiency into account. Therefore, an existing CCV model was adapted to the improved burn rate model, as will be described in more detail in Chapter 4.7. If the influence of CCV on the HRR is to be modeled correctly, the τ_L-approach would have to be made CCV-dependent. However, this subject is not within the scope of this work.

An alternative to using the τ_L-approach would be to reliably simulate various SWC at one operating point. These SWC can then be averaged to obtain the AWC. This would either require a sophisticated averaging method of only a few SWC or the detailed modeling of numerous SWC matching all different HRR characteristics caused by CCV. The latter would require detailed knowledge of local effects like inhomogeneities of the temperature or mixture composition and the local turbulence level. These values are not yet available in the 0D/1D model class, but current research projects (see [50]) and planned follow-up projects work on that issue and might lay the foundation to follow this approach in the future.

Nevertheless, it is to be expected that the local TKE distribution will already influence the simulation when using the τ_L-approach. For example, it could lead to a reduced flame speed towards the end of combustion. In general terms, taking into account the TKE distribution, fewer influences have to be compensated by the τ_L-approach. When taking a closer look at the conditions in the combustion chamber, a very small distance between flame and wall can be observed at only 50 % mass fraction burnt, which corresponds to a burnt volume of about 75 %. With a cylindrical simplification of the combustion chamber of engine A, for example, the distance is 6 mm, assuming a perfectly spherical flame propagation (see Figure 4.15). Thus it becomes clear that wall effects can have an influence already at relatively low mass fractions burnt, especially when considering an off-center spark plug position and non-uniform flame propagation.

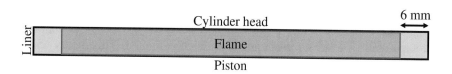

Figure 4.15: Schematic flame propagation at MFB50 and $V_{b,rel} = 75\,\%$, engine B

4.5 Influence of the Local Turbulent Kinetic Energy

To investigate the local TKE distribution, 3D CFD calculations from [50] were used. The distribution of TKE at TDC is depicted in Figure 4.16. Figure 4.17 suggests that the TKE distribution is – relative to the mass-averaged mean value – not dependent on the engine speed. An approach to consider the local TKE in the QD burn rate model presented in this study could be to impose a TKE correction as a function of the flame radius. This correction could be obtained from 3D CFD results by calculating the mean TKE for a series of spherical shells with constant thickness and increasing radius. However, these results are dependent on the thickness of the shell, which can be interpreted as a moving average that smooths the local TKE distribution. Furthermore, it is generally known that the flame does not propagate in a perfectly spherical way. This is supported by the movement of the flame centroid mentioned in Chapter 4.4. Nevertheless, the approach of imposing a flame-radius-dependent TKE correction is a first step towards considering local effects in the 0D/QD model class.

When comparing the relative change in TKE with flame radius, a dependency on piston position is observed. Figure 4.18 shows that by replacing the absolute flame radius with a relative radius (eq. 4.14), this dependency disappears. Similar observations were made by *Malcher et al.* [95] , who took the relative burnt volume as reference and integrated the local TKE into the burn rate model by means of a look-up table.

Since not the TKE, but the turbulent fluctuation velocity u' is used to calculate s_T, the change in TKE is transferred to a change in u' using the relationship $u' = \sqrt{2/3 \cdot \text{TKE}}$ in Figure 4.18. In consequence, the change in u' with relative flame radius can easily be modeled. For this, the mathematical approach published by *Vibe* [141] for heat release rates was taken for the purpose of approximation:

$$u'_{\text{fac}} = c \cdot a \cdot (m+1) \cdot r_{\text{rel}}^m \cdot e^{-a \cdot r_{\text{rel}}^{m+1}} \qquad \text{eq. 4.13}$$

$$r_{\text{rel}} = \frac{r_{\text{flame}}}{r_{\text{max}}} \qquad \text{eq. 4.14}$$

$$r_{\text{max}}(CAD) = \sqrt{\left(\frac{V_{\text{d}}(CAD)}{A_{\text{piston}}} + TDC_{\text{clearance}}\right)^2 + r_{\text{bore}}^2} \qquad \text{eq. 4.15}$$

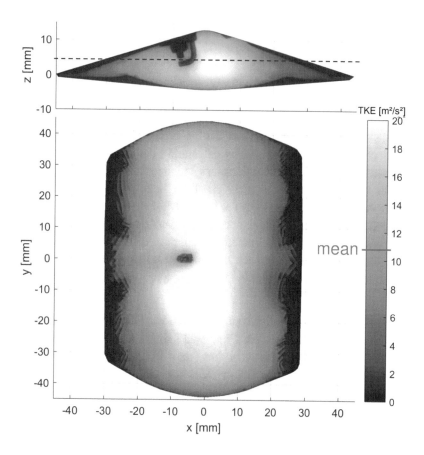

Figure 4.16: TKE distribution at TDC, engine B, 2000 rpm [72]

$$c = \frac{max\left(u'_{\text{fac}}\right), target}{max\left(u'_{\text{fac}}\right)} \qquad \text{eq. 4.16}$$

The relative crank angle of *Vibe* is replaced by the relative flame radius. The maximum flame radius is a function of CAD and is calculated on the basis of simplified geometric considerations. $TDC_{clearance}$ allows to account for the distance between piston and spark plug. In a first approach, $TDC_{clearance}$ can be calculated based on the compression volume and the cylinder bore. The baseline *Vibe* calibration parameters a and m are now used to match the shape

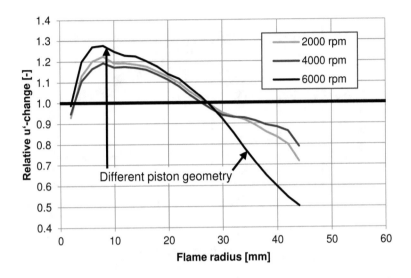

Figure 4.17: Influence of engine speed and piston geometry on relative u'-change with flame radius, engine B [72]

of the relative u'-change. The parameter c is the replacement for the total heat released. Here, it is calculated on the basis of a maximum u'-factor $(max(u'_{fac}), target)$, which can be set by the user. As highlighted in Figure 4.18, this model approach allows a high quality fit of u'-changes from 3D CFD calculations. When calibrating a and m, it is important to maintain a plausible trend of u'_{fac}. For example, a permanent increase in u'_{fac} with flame radius should be avoided. Furthermore, based on the 3D CFD results, an overall influence of ±40 % to ±50 % should not be exceeded. However, this depends on the combustion chamber geometry and the flow characteristics. Due to the uncertainties of determining the effective local TKE, as mentioned above, a calibratable model seems to be more advantageous than a tabulated TKE.

The distribution of TKE and u', illustrated in Figure 4.16 and Figure 4.18 respectively, also shows an increase in turbulence level and, with that, in turbulent flame speed at small flame radii. In the literature, such as in [140], it was often proposed to employ an extra model for this early stage of combustion.

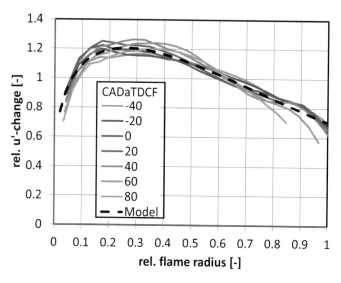

Figure 4.18: Comparison of relative u'-change at different CAD with the modeling approach, engine B [72]

The necessity for such a model was usually explained with an increasing spectrum of turbulence acting upon the growing flame kernel. This explanation seems unlikely when relating l_{int} of about 1 mm to a typical air gap between the spark plug electrodes of 0.6 mm. In addition, the thermal expansion due to combustion causes the flame kernel to exceed the largest scale of turbulence even faster. In contrast, the available time for the turbulence to wrinkle the flame might be too short [20], therefore supporting the original explanation. Nevertheless, considering Figure 4.16 and Figure 4.18, it seems to be more likely that the general increase in turbulence level at small flame radii is the reason for a slower start of combustion, instead of the reduced interaction of turbulence with the small flame kernel. Furthermore, a difference between simulated and measured HRR of an AWC at very low MFB is often used to justify the utilization of an extra model for the early combustion stage. However, Figure 4.14 shows that by averaging multiple HRR of SWC, a gentle combustion start is achieved in the mean HRR, despite a very fast start of combustion at the HRR of SWC. In consequence, a slow combustion start could be caused by CCV differences of the SWC instead of physical effects at small flame radii.

Furthermore, the TKE distribution indicates that in the example illustrated in Figure 4.16, the flame will propagate faster towards the center of the combustion chamber due to higher TKE. This not only leads to a non-uniform flame propagation, but also shifts the flame center towards the combustion chamber center, despite an off-center spark plug. Again, the use of the spark plug position as a calibration parameter in simulation is thereby justified.

4.6 Overview: Improved Burn Rate Model

In summary, the baseline burn rate model was improved by

1. investigating and disproving the influence of fuel on flame wrinkling

2. modeling the laminar flame speed and thickness for different fuels on the basis of reaction kinetics calculations

3. using a more sophisticated model for the turbulent flame speed s_T

4. improving the calculation of the characteristic burn-up time τ_L

5. introducing an approach to account for the local u'.

Despite a difference between the results of different reaction mechanisms, reaction kinetics calculations proved to be a reliable source for s_L and δ_L. The s_T-model by *Peters* showed the most plausible reaction to boundary condition changes. According to [140], it is reasonable to use the turbulent flame speed s_T as entrainment velocity u_E, if the s_T-definition is appropriate. As $s_{T,\text{Peters}}$ is defined as leading edge speed, this criterion is met. Consequently, the baseline u_E (eq. 2.5) was replaced with $s_{T,\text{Peters}}$ in eq. 4.21. The approach to incorporate the local u' promises to improve the shape of the HRR, which will be validated later. The improved calculation of τ_L not only allows a proper prediction of boundary conditions changes, but is also reasonable from a phenomenological and mathematical point of view. The model consists of the following equations:

$$dQ_b = dm_b \cdot H_{u,\text{mix,grav.}} \qquad \text{eq. 4.17}$$

$$\frac{dm_b}{dt} = \frac{m_F}{\tau_L} = \frac{m_E - m_b}{\tau_L} \qquad \text{eq. 4.18}$$

$$\tau_L = a_\tau \cdot \frac{l_{int}}{u_E} \qquad\qquad \text{eq. 4.19}$$

$$\frac{dm_E}{dt} = \rho_u \cdot A_{fl} \cdot u_E \qquad\qquad \text{eq. 4.20}$$

$$u_E = a_u \cdot s_L \cdot \left(1 + 0.195 \cdot \frac{l_{int}}{\delta_L} \cdot \left(\sqrt{1 + \frac{20.5}{\frac{s_L}{u'_{eff}} \cdot \frac{l_{int}}{\delta_L}}} - 1\right)\right) \qquad\qquad \text{eq. 4.21}$$

$$u'_{eff} = u'_{fac}(r_{flame}) \cdot u' \qquad\qquad \text{eq. 4.22}$$

$$l_{int} = a_{l_{int}} \cdot \left(\frac{6}{\pi} \cdot V_{cyl}(CAD)\right)^{\frac{1}{3}} \qquad\qquad \text{eq. 4.23}$$

The laminar flame speed s_L and laminar flame thickness δ_L are calculated using the models developed in Chapter 4.1 and Chapter 4.3. The turbulent fluctuation velocity u' is calculated using the turbulence models published in [15] and [62] (see Chapter 2.2). If available, u' from 3D CFD calculations can also be imported. The model to consider the local u' via u'_{fac} was developed in Chapter 4.5. $a_{l_{int}}$ is needed to scale down the integral length scale l_{int} to the level of 3D CFD results, as presented in Figure 4.19. Without scaling, the calculated l_{int} would be about a factor of 45 higher than the results obtained with StarCCM+. This would cause implausible values for u_E. Interesting to see is the difference in level and especially trend between the results of StarCCM+ and Ansys Forte. The trend shown by Ansys Forte was also observed in [149] for a Diesel-type combustion chamber. Since it was not possible to determine which result is correct, the possibility to import the integral length scale was included in addition to using the model approach.

a_τ and a_u are calibration factors. Without using the τ_L-approach (compare Figure 4.11), a_u needs to be set to a value of about 1.7. This approach changes the model, so that $dm_b = dm_E$. Considering the definition of the turbulent flame speed already mentioned in Chapter 2.4 and [18], this model assumption is linked to a turbulent flame speed defined in the middle of the trailing edge and the leading edge of the flame. However, the turbulent flame speed model of *Peters* is defined at the leading edge of the flame. According to [19], the different flame speed definitions can be converted into each other. For a fully developed flame, the conversion factor is about 1.58 [19] for the change in

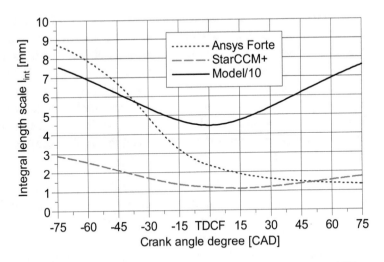

Figure 4.19: Modeled integral length scale vs. 3D CFD results [72]

definition considered here. This could explain the need for the calibration factor a_u and its value of 1.7. Furthermore, a_u covers, for example, errors in the s_T-model itself, the level of u' and the uncertainty in the level of the laminar flame thickness due to different definitions and the underlying thermodynamic and transport data of the reaction mechanism. When using the τ_L-approach, the calibration factor a_u is of the order of 3 and necessary to "feed" the flame front to compensate for the delay caused by the characteristic burn-up time. As mentioned above, this delay is needed to smooth the shape of the HRR. The smoothing effect is changed by varying the calibration parameter a_τ. a_τ is in the order of 12 and should lead to a characteristic burn-up time in the order of 1 ms.

Furthermore, the flame surface area needs to be calculated. At best, it is tabulated on the basis of the detailed combustion chamber geometry in a CAD software. Alternatively, it is shown in Appendix A7.1 that by simplifying the detailed combustion chamber with a cylindrical geometry, similar HRR shapes can be obtained. For this, the spark plug position needs to be changed to match the results of the detailed combustion chamber geometry, with the horizontal spark plug position having the major influence.

In conclusion, no calibration parameters would be needed if the integral length scale is of the proper magnitude, the flame surface area is known from detailed 3D combustion chamber data and the τ_L-approach is not used. Considering the theory of averaging multiple SWC from simulation to obtain the HRR of an AWC (see 4.4), dismissing the τ_L-approach seems possible. As already described above, information of numerous local effects need to be known to follow this idea, but is not yet available in the 0D/1D model class. Therefore, the τ_L-approach is needed to cover those effects and to calculate the HRR of an AWC in simulation, for the "price" of two calibration parameters.

In the typical case, neither u', nor the flame area, nor the integral length scale are given. This increases the calibration parameters from two (a_τ, a_u) to the worst case maximum of seven: four (a_τ, a_u, $a_{l_{int}}$ and the horizontal spark plug position sp_{hor}) for the burn rate model and three (a, m and $max(u'_{fac}), target$) for the fine-tuning via local u'. The best way to deal with the different calibration parameters will be explained in Chapter 5.8. But first, the linking of the improved burn rate model to an existing CCV model is described. This is important since, as stated above, the burn rate model calculates HRR which are only valid for a certain CCV level. The influence of increasing CCV (e.g. the reduction of engine efficiency) needs to be covered by a separate model.

4.7 Linking to a Cycle-to-Cycle Variations Model

In order to account for CCV influence, a CCV model, for example, the one developed in [146] and [147], can be used. This is based on the baseline burn rate model described in Chapter 2.2. The CCV model applies stochastic noise of constant bandwidth to turbulence (represented by a variation of combustion start, calibrated by the parameter φ_{ZS}) and mixture distribution (represented by a change in laminar flame speed, calibrated by the parameter χ_{ZS}). Depending on engine operation conditions like charge dilution, engine speed or MFB50, the noise causes fluctuations in sub-models, which eventually result in more or less severe fluctuations of the indicated mean effective pressure (IMEP). This allows the prediction of statistical IMEP fluctuations, represented by the coefficient of variation (COV$_{IMEP}$). If the COV$_{IMEP}$ exceeds a certain limit, the

CCV are considered too high for stable engine operation. In [147], the model was validated for three different passenger car engines fueled with gasoline and a MTU 4000 series natural gas engine with 57.2 l displacement volume. The model allowed a reliable prediction of the COV_{IMEP} for all engines.

With the improvement of the burn rate model, the influence of χ_{ZS} on the burn rate model was changed, too. As stated above, the baseline burn rate model overestimates the s_L-influence on the HRR via the τ_L-approach. For the CCV model, this leads to a strong change of HRR when changing s_L via χ_{ZS}. To maintain this strong influence, the CCV model varies u_E instead of s_L when using the improved burn rate model. A variation of s_L might still be possible. However, the reduced influence of s_L could necessitate an excessive variation, possibly exceeding flammability limits. From a phenomenological point of view, the variation of u_E instead of s_L is also reasonable (see eq. 4.21): due to CCV, not only s_L, but also δ_L and the effective (local) u' vary, consequently changing u_E. The reliable model performance in combination with the improved burn rate model was proven for a large gas engine by the author in [73]. The most important validation results will be reproduced in Chapter 5.4.

5 Burn Rate Model Validation

The improved burn rate model was validated using engines A and B. Their data are provided in Table 3.1 and Table 3.2. The PTA settings were given in Chapter 3. Additionally, measurements of engines C, D, E and F were available for model validation. Engine C is a high-turbulence long-stroke engine with different pistons to change the compression ratio. Engine D allowed the comparison of gasoline to DMC+. The influence of water injection was investigated on engine E. Engine F is a large gas engine with a prechamber spark plug.

For engines C, D, E and F, the same PTA settings as for engines A and B were used. Due to a lack of mandatory data to set up an engine model in GT-Power, the PTA for engines E and F was performed without gas exchange. This means that the calculation started at intake valve closing and ended at exhaust valve opening. For this, the internal residual gas fraction had to be estimated. Since both engines feature a small valve overlap, this value was set to 3 %. Furthermore, not enough data was available to model the heat transfer from the cylinder walls to the coolant. For this reason, a constant cylinder wall temperature was imposed for engines E and F. While this assumption is valid for engine E, the wider variation of boundary conditions on engine F would cause changing wall temperatures. From analyzing the wall temperature change of engines A, B and C, a maximum error of ±10 K was estimated. This range was supported by a simulative investigation of the wall temperature change on a virtual engine for varying MFB50 at constant IMEP. Due to the high temperature difference between the mean gas temperature and the cylinder walls, the wall temperature uncertainty lead to an IMEP error of only ±0.25 %.

Besides the general quality of the model, its validation will allow to eliminate specific uncertainties. Concerning the laminar flame speed, this is the influence of boundary conditions like temperature, pressure, λ or the fuel. Chapter 4.1.5 already proved that the laminar flame speed model can cover those influences reasonably well. However, they are dependent on the chosen reaction mechanism (see Figure 2.7). The mechanism published by *Cai et al.* [22] was

© The Author(s), under exclusive license to
Springer Fachmedien Wiesbaden GmbH, part of Springer Nature 2021
S. Hann, *A Quasi-Dimensional SI Burn Rate Model for Carbon-Neutral Fuels*,
Wissenschaftliche Reihe Fahrzeugtechnik Universität Stuttgart,

chosen as a starting point due to its wide validation range. Nevertheless, its characteristics concerning boundary condition changes still need to be evaluated. Another uncertainty is depicted in Figure 4.7, highlighting that different turbulent flame speed models react differently to changes in the turbulent fluctuation velocity u', the laminar flame speed and the pressure. While the influence of pressure and changing laminar flame speed with λ on the turbulent flame speed model of *Peters* were validated in Chapter 4.2 and Chapter 4.11, respectively, the influence of u' has not been tested yet. In Chapter 3, a fuel influence on flame wrinkling was already discarded based on a measurement data analysis. However, a further validation of this conclusion is necessary.

From the experience made during the validation process, a guideline to calibrate the model was derived and will be given in the last subchapter. Typical values and plausible ranges of all model calibration parameters will be provided to assist in the application of the burn rate model to new engines.

5.1 Engine A

The measurements carried out on engine A represent the main basis of this study. They were planned in cooperation with and performed by Marcel Eberbach [38] in the course of the FVV project #1213 "Methane Fuels II". Gasoline, ethanol and methane were used as fuels and injected into the intake port. For all fuels, λ was varied at constant energy input, resulting in an almost constant IMEP of 11 bar. The engine speed was set to 1500 rpm, 2000 rpm and 3000 rpm for each variation of λ. These operating conditions allowed to validate the chosen reaction mechanism, the influence of fuel on flame wrinkling and the change in s_T with u'.

The intake air temperature was set to 25 °C in order to account for the low knock resistance of ethanol and especially gasoline. With that, it was possible to operate the engine at MFB50 = 8 CAD and $\lambda = 1$. Due to the increase in boost pressure for leaner mixtures, the knock tendency of gasoline increased, which made a late-shifting of MFB50 necessary. With it, the COV_{IMEP} increased as well, as illustrated in Figure 3.3. Despite a constant MFB50, ethanol also showed a high COV_{IMEP}. This was most likely caused by the formation of a wall film in the intake port due to the low intake air temperature and the high evaporation enthalpy of ethanol. In order to reduce the influence of CCV on the averaged working cycle (AWC), bad or not burning single working cycles (SWC) were not considered in the calculation of the AWC.

The burn rate model was calibrated for methane at 2000 rpm and $\lambda = 1$. The turbulence was calculated using the model described in [15]. The approach to account for the local u'-distribution was not used. The integral length scale was scaled down by $a_{l_{int}} = 0.022$ in accordance with Figure 4.19. The combustion chamber was replicated in 3D from pictures in order to calculate the flame surface based on the detailed geometry. The optimal model calibration parameters a_τ and a_u were determined by performing a parameter variation, leading to $a_\tau = 12$ and $a_u = 2.75$. The interpretation of these values was already given above. This model calibration was kept constant for a variation of fuel, λ and engine speed. The simulated IMEP and HRR are compared with results from PTA in Figure 5.1 to Figure 5.4.

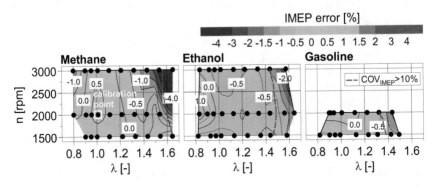

Figure 5.1: IMEP error of pure fuels: PTA vs. simulation, engine A [72] (updated and expanded)

In contrast to Figure 4.10 (baseline model), the IMEP error did not change with a variation of λ. This proves the validity of the new τ_L-approach. The heat release rate change for different λ, summarized in Figure 5.2 for methane, underlines this conclusion: while the improved model predicted the change of HRR well, the HRR error of the baseline model followed the IMEP error given in Figure 4.10. Here, $\chi_{Taylor} = 7$ was used as reference. For smaller χ_{Taylor}, the

IMEP prediction quality would have been higher, but then the shape of the heat release would have become more and more similar to those plotted in Figure 4.11. The good predictive ability of the improved model also indicates that the change in laminar flame speed and thickness within this model structure is plausible and the reaction mechanism published by *Cai et al.* gives plausible results for methane at changing boundary conditions. With methane having a *Lewis* number of one, fuel-dependent flame wrinkling cannot distort these conclusions.

Figure 5.1 illustrates that the change in engine speed was also predicted well. This shows that the improved burn rate model, and with that, the *Peters* model for s_T, react properly to changes in the turbulence level. At very lean λ, the error increased. As represented by the broken line in Figure 5.1, the COV_{IMEP} was above 10 % and reached a maximum of 25 % (see Figure 3.3) at lean conditions. Despite the reduction of the influence of CCV on HRR by dismissing bad and not burning SWC, the reliability of an AWC at such high COV_{IMEP} values is questionable. Usually, $COV_{IMEP} = 3$ % is defined as an acceptable limit of stable engine operation. Since engine A is a prototype research engine, all operating points were considered to highlight the influence of CCV on HRR and the limitations of the model.

When the fuel was switched to ethanol, the prediction quality of the model did not change. This is highlighted in Figure 5.1 for a change in IMEP error and in Figure 5.3 for a change of HRR. It thus can be concluded that a fuel influence on flame wrinkling does not need to be taken into account and the change in fuel is sufficiently modeled by changing the laminar flame speed and flame thickness. This is underlined by the good prediction of leaner λ, where a strong influence of fuel-dependent flame wrinkling was expected from Figure 2.12. The error at the leanest λ is most likely caused by the high COV_{IMEP} mentioned above. The validation using engine C proves that the model works well even at $\lambda = 1.8$, if the CCV are low. Again, the *Cai* mechanism gives plausible s_L-values, but the results of different reaction mechanisms are similar for ethanol anyway, as summarized in Figure 2.7. The baseline model overestimated the HRR change with changing fuel and λ. This is a consequence of the excessive s_L-influence via the τ_L-approach. Due to the strong difference between the laminar flame speed of methane and ethanol (see Appendix A1.1), the model error is very prominent here.

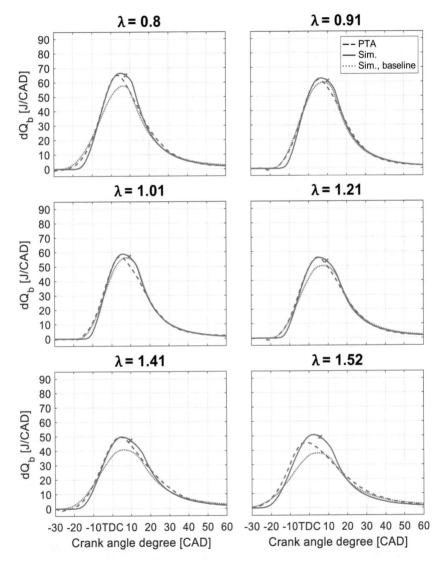

Figure 5.2: HRR change with λ for methane, engine A, 2000 rpm [72] (recalculated and expanded)

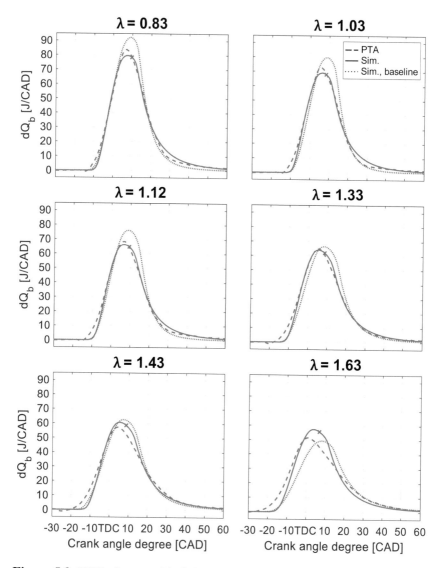

Figure 5.3: HRR change with λ for ethanol, engine A, 2000 rpm [72] (recalculated and expanded)

The proper prediction of gasoline at stoichiometric and lean conditions, illustrated in Figure 5.1 and Figure 5.4, further confirms the lack of a fuel-dependent flame wrinkling. From Figure 2.12, not only an offset of the turbulent flame speed, but also a much stronger change when varying λ was expected. The improved burn rate model covered these variations reasonably well by only accounting for changes in the laminar flame speed and the laminar flame thickness. At rich conditions, the HRR was underestimated. This only led to a low IMEP error and a peak pressure error of 0.5 %. Since a variation of fuel as well as λ for different fuels were predicted well, a fuel influence on flame wrinkling as reason for the underestimation of the HRR is unlikely. Figure 2.7 shows that the reaction mechanism published by *Liu et al.* [93] gives a higher laminar flame speed at rich conditions. When this s_L was tested in simulation by an upscaling of the *Cai* results by 10 %, the HRR was increased, but not significantly (see Figure 5.5). With that, an error in s_L could only be a small part of the reason. Another possible explanation might be an error in the measurement of λ. As mentioned above, the formation of a wall film in the intake ports was observed. This caused a fluctuation in the measured λ. At rich conditions, the formation of the wall film should be strongest and should therefore have the strongest influence. However, no definitive reason could be found for the slight underestimation of HRR. At very lean conditions, the measured HRR is again influenced by the high COV_{IMEP} and thus considered unreliable. As with ethanol, the baseline model overestimated the fuel and λ influence on HRR.

Figure 5.6 depicts the change in IMEP error for the admixture of both ethane and hydrogen to methane. The laminar flame speed and the laminar flame thickness of the fuel mixtures were modeled by performing a linear interpolation, related to the mass fraction of ethane or hydrogen, respectively. This approach was already introduced in Chapter 4.1.3 for the laminar flame speed and in Chapter 4.3 for the laminar flame thickness. The low change in IMEP error with an increase in the fraction of the respective secondary component underlines the applicability of this method. In contrast to Chapter 4.3, the modeled instead of the calculated laminar flame speed was now used as an input parameter for the laminar flame thickness model. Due to the error propagation, the low error of both models ensures a reliable prediction of the fuel influence on IMEP.

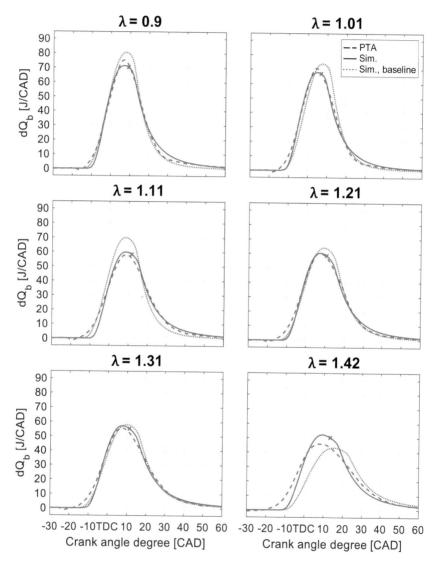

Figure 5.4: HRR change with λ for gasoline, engine A, 2000 rpm [72] (recalculated and expanded)

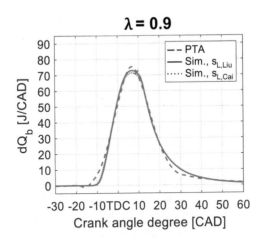

Figure 5.5: HRR change with s_L from the *Liu* mechanism (10 % upscaling of *Cai* results), gasoline, 2000 rpm, engine A [72]

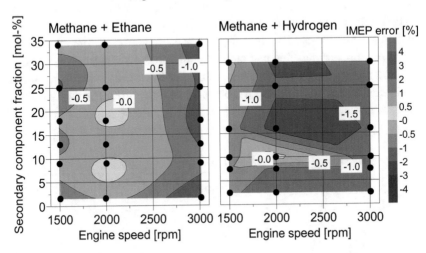

Figure 5.6: IMEP error of fuel mixtures: PTA vs. simulation, engine A

Figures 5.7 and 5.8 summarize the change of HRR with increasing amounts of ethane or hydrogen at different engine speeds. The measurements underlying Figures 5.6, 5.7 and 5.8 were performed on the same engine, but five years before the investigation of pure fuels happened. A test bench fire occurring during this period resulted in the engine having to be rebuilt. In consequence, the calibration parameter a_u had to be increased from 2.75 to 2.85 in order to match the measurement results. As this value was kept constant for the variation of fuel mixture composition, Figures 5.6, 5.7 and 5.8 nevertheless highlight the ability of the approach to account for changes in fuel composition by a mass fraction-based, linear interpolation to properly predict these changes in engine simulation. Only the HRR of methane + 2.5 mol−% H_2 was slightly overestimated. This underlines the conclusion that the modeling of changes in s_L and δ_L is sufficient to cover the fuel influence on HRR in engine simulation, as was already mentioned above. Considering the modeling of a CNG substitute, as suggested in Chapter 4.1.2, Figure 5.7 illustrates the sufficiency to only account for ethane equivalents above 6 mol−%. Below 6 mol−%, the change in HRR will be small, but the model provides the necessary accuracy for such detailed investigations. Due to a lack of measurement data, the model quality for mixtures of gasoline and ethanol could not be validated. Considering the good results for the admixture of ethane or hydrogen to methane and the similar quality of the flame speed models for fuel mixtures (Figure 4.6), a reliable prediction of HRR changes due to ethanol admixture can still be expected, especially regarding its lower flame speed increase (see Figure 4.4).

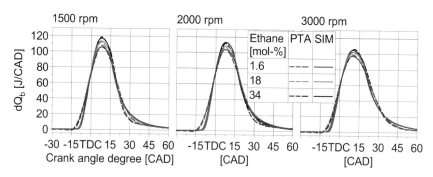

Figure 5.7: HRR change with fuel composition: methane + ethane, engine A, IMEP ≈ 20 bar, variable MFB50

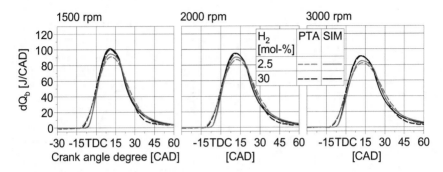

Figure 5.8: HRR change with fuel composition: methane + hydrogen, engine A, IMEP ≈ 18 bar, variable MFB50

In conclusion, the model can predict changes in fuel, λ and engine speed with a constant set of calibration parameters. This shows that a fuel influence on flame wrinkling is unlikely. Furthermore, the new approach to calculate τ_L performs well. In the simulation environment of the burn rate model, s_L calculated with the reaction mechanism of *Cai et al.* [22] seems plausible and the s_L-model proves its high quality. This was further investigated by importing a table of s_L instead of using the model. For operating points with a reasonable COV$_{IMEP}$, a maximum change in IMEP error of 0.2 percentage points was observed and underlines the good quality of the s_L- and δ_L-model.

One prominent feature of almost all the simulated HRR was the wider maximum. It occurred usually at about MFB50 (marked by the cross on the solid line in Figure 5.2 to Figure 5.4), which corresponds to $V_{b,rel} = 75\,\%$. There, the distance between an unwrinkled spherical flame and the cylinder walls is only a few millimeters, as stated in Chapter 4.4. Considering the wrinkled and deformed shape of the flame in addition to its non-uniform propagation, contact between flame and cylinder wall is likely. Therefore, the influence of using the local u' (see Chapter 4.5) is investigated in Figure 5.9. Here, the change in u' given in Figure 4.18 was used. The small increase and decrease in u' of ±20 % led to a HRR which was in better agreement with PTA results. With that, it is reasonable to account for the local change in u', since a wall contact and thus a reduced u' is plausible and results in plausible HRR changes. Furthermore, it reduces the phenomena which need to be covered by the τ_L-approach.

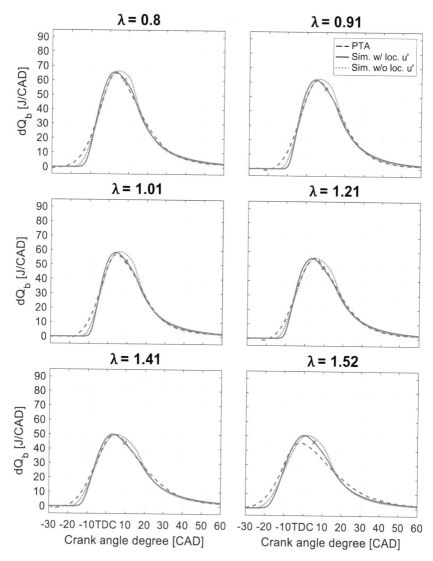

Figure 5.9: Influence of local u' on the HRR of methane, engine A, 2000 rpm
[72] (recalculated and expanded)

5.2 Engine B

The technical data of engine B are given in Table 3.2. The measurements were kindly provided by Dr. David Lejsek (Robert Bosch GmbH) to be used in the FVV project #1213 "Methane Fuels II". In Chapter 3.1, these data were employed to investigate the possible influence of the *Darrieus-Landau* instability on engine combustion. Here, the data allow a further validation of the improved burn rate model. As the whole engine map was measured for methane and gasoline, again using port fuel injection, not only the prediction of changing fuels, but also the prediction of a wider variation of engine speed and especially pressure (by varying the engine load) could be validated. Furthermore, from 3D CFD-calculations performed in [50], the turbulent kinetic energy (see exemplarily Figure 4.16) as well as the integral length scale were available. While the turbulent kinetic energy was imported directly in the burn rate model, the integral length scale calculated in the burn rate model was scaled down to match the StarCCM+ results (see Figure 4.19) by setting $a_{l_{int}}$ to 0.022. Additionally, the detailed flame surface area could be calculated from available 3D data. With that, several uncertainties could be removed. In consequence, only the parameters a_τ and a_u needed to be used for the model calibration.

For this engine, the model was calibrated for methane at a medium load of 12 bar IMEP and a medium engine speed of 4000 rpm. In order to match the IMEP and HRR, a_τ and a_u were set to 12 and 3.6, respectively. Compared to engine A, only a_u was changed. Based on this calibration, the engine maps of both cylinders and both fuels were simulated without recalibration. The IMEP error is shown in Figure 5.10. In a wide range of boundary conditions, the IMEP error was below ±1 %. Even at late MFB50 of about 22 CAD (circle in Figure 5.10) or MFB50 = 17 CAD in combination with $\lambda = 0.8$ (square in Figure 5.10), the error did not increase significantly. Furthermore, the influence of importing s_L and δ_L instead of using the models was investigated. The resulting change in IMEP error of only ±0.25 percentage points again proves the quality of the models for s_L and δ_L.

Figure 5.11 summarizes the HRR of different engine loads at different engine speeds for both fuels. Again, the change in boundary conditions can be predicted well when using the improved burn rate model. This is another proof of the negligible influence of fuel on flame wrinkling. The proper reaction to

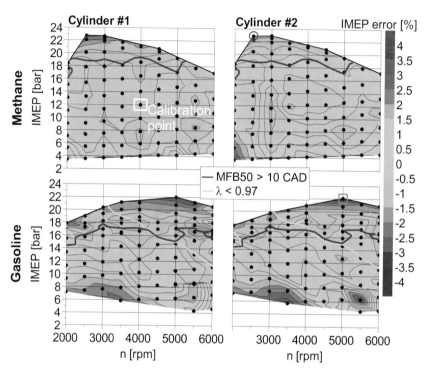

Figure 5.10: IMEP error, PTA vs. simulation, engine B [72] (recalculated)

changes in engine speed and load proves the reliable reaction of the model to changes in turbulence, temperature and pressure. Furthermore, it shows that an influence of the DL instability on combustion is unlikely, as elaborated in more detail in Chapter 3.1. In Figure 5.12, a variation of load and MFB50 up to 21 CAD is given for 2500 rpm. The good quality of the HRR prediction highlights the proper model reaction to MFB50 changes, where not only the turbulence level is changing, but also the flame geometry.

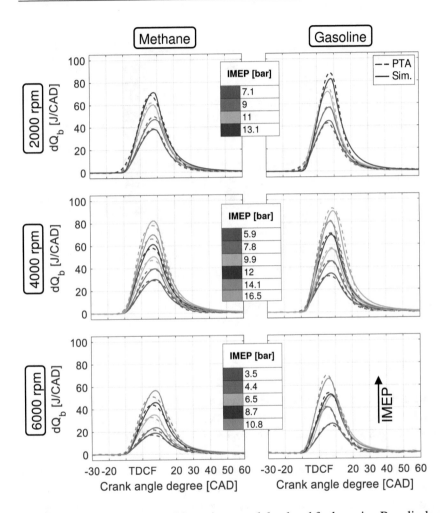

Figure 5.11: HRR change with engine speed, load and fuel, engine B, cylinder
1 [72] (recalculated)

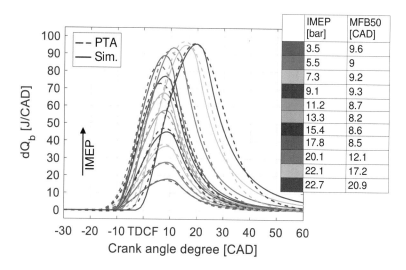

Figure 5.12: HRR change with engine load and MFB50, 2500 rpm, methane, engine B, cylinder 1 [72] (recalculated)

5.3 Engine C

The measurement data of engine C were a result of the work done in the FVV project #1307 "ICE2025+: Ultimate System Efficiency" and were kindly provided by the RWTH Aachen with agreement of Mr. Arndt Döhler (Opel Automobile GmbH), the chairmen of the project. The PTA and the burn rate model calibration were performed by *Crönert* in [28] under the supervision of the author. The following investigation was already published by the author in [72]. Due to the updated laminar flame speed model, all simulations were repeated for this thesis. The technical data of engine C and its operating conditions are given in Table 5.1. For the measurements investigated, only DI was used. The design target of the cylinder head was to reach a high level of turbulence, additionally increased by the long stroke configuration. The compression ratio was changed by using different pistons.

The burn rate model was calibrated for methanol at $\lambda = 1$. The flame surface area was calculated from 3D data on the basis of a remodeled geometry using

Table 5.1: Engine C: technical data and operating conditions

Cylinder no.	1
CR [-]	13 (gasoline) and 14.7 (methanol & MeFo)
Bore [mm]	75
Stroke [mm]	90.5
IMAT [°C]	30
Load (IMEP [bar])	methanol, MeFo: 16, gasoline: 12
Engine speed [rpm]	methanol, MeFo: 2500, gasoline: 2000
Variation	λ and EGR up to misfire limit
Injection	DI and PFI

combustion chamber pictures. The turbulence was calculated using the model described in [15]. The plausibility of the turbulence was checked by comparing the model results to 3D CFD results from engine B. This was done to definitely cover the high turbulence nature of the engine. For the integral length scale, the modeling approach shown in Figure 4.19 was used.

In a first attempt, the model was only slightly re-calibrated. a_u was set to 3, $a_{l_{int}}$ to 0.022, with the local TKE switched off. These calibration parameter values are in the range of engine A and B, whose turbulence level is much lower. a_τ was reduced to 8 due to the very low CCV given in Figure 5.13. As written above, the τ_L-approach resembles a method to account for CCV influences on the HRR. Therefore, it is reasonable to reduce a_τ and with that, the characteristic burn-up time. With this calibration, an IMEP error very similar to the one illustrated in Figure 5.13 was achieved. The burn rate model thus reacts properly to a change in turbulence and gives plausible results even at very high turbulence levels. Furthermore, despite a significant difference between the characteristics of engine A, B and C, a similar set of calibration parameters allowed to predict the IMEP. This underlines the validity of the burn rate model assumptions as well as the reasons for the need of calibration parameters given in Chapter 4.6. However, with this set of parameters, the change of HRR with λ for methanol was not matched perfectly.

This was achieved by reducing $a_{l_{int}}$ to 0.0063, which led to a minimum integral length scale of about 0.25 mm. Since the integral length scale not only

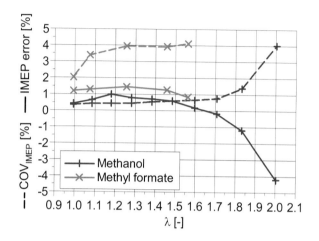

Figure 5.13: IMEP error and COV_{IMEP} of MeFo and methanol, engine C, 2500 rpm [72] (recalculated)

influences u_E, but also τ_L, a_τ needed to be increased to 25 to compensate for the small integral length scale. The magnitude of the characteristic burn-up time thereby remained at a similar level as with the first calibration approach. Considering the very high compression ratio of 14.7 and the resulting, small dimensions of the combustion chamber, a reduction of the integral length scale seems plausible. Further fine-tuning was done by using the correction for local turbulence ($a = 0.6$, $m = 0.1$, $max(u'_{fac})$, $target = 1.4$). Again, due to the shape of the combustion chamber (bowl in the piston center, surrounded by large squish areas), a strong reduction of turbulence due to increased wall friction is to be expected for larger flame radii, consequently increasing the relevance of the local TKE distribution. This assumption is underlined by the piston geometry influence on relative u'-change highlighted in Figure 4.17. By using the re-calibration and fine-tuning described above, the IMEP error illustrated in Figure 5.13 and the HRR given in Figure 5.14 were achieved.

The measured heat release rates displayed in Figure 5.14 show, especially at leaner λ, a deformation of the peak, which did not occur on the other engines investigated. One reason might be the pronounced piston bowl, leading to an intermediate reduction of flame surface. Furthermore, the large squish areas might cause an increase in turbulence shortly after TDC by generating a squish

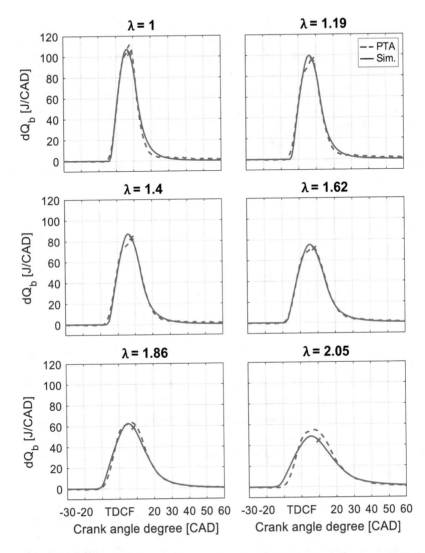

Figure 5.14: HRR change with λ for methanol, engine C, 2500 rpm [72] (re-calculated)

flow. The very low level of COV_{IMEP} of this engine also leads to a low smoothing effect of CCV on the HRR shape, as discussed in Chapter 4.4. With that, the average HRR is more similar to those of rugged SWC. All these (local) effects cannot (fully) be covered in the 0D/1D model class. Consequently, the simulated HRR does not show the deformation. Nevertheless, the influence of λ was predicted well, again highlighting the proper reaction of the burn rate model to changes in s_L. As with methane, this conclusion is not affected by a possible fuel influence on flame wrinkling, since the *Lewis* number of methanol is close to one (see Figure 2.12). The good prediction at high values of λ also indicates that for engine A, the high CCV level does influence the change of HRR with λ there. From the quality of the results, it can be inferred that it seems to be sufficient to only change the laminar flame speed and thickness to account for the fuel influence on burn rate modeling. Especially when adding new fuels to the model, this is important knowledge.

Figure 5.15 summarizes the HRR simulated for methyl formate (MeFo) when using the same refined model calibration as for methanol. Despite some small HRR differences between simulation and measurement, the IMEP error (see Figure 5.13) did not change with λ. This again proves that the fuel influence can be predicted properly. Furthermore, the measurements of MeFo were influenced by a ruptured fuel control valve sealing, which caused fluctuations in the fuel pressure. Figure 5.13 illustrates the resulting increase in CCV. Furthermore, this might also have influenced the measured λ-value of the AWC. Considering these uncertainties, the prediction quality of the model is good, especially for the large change in laminar flame speed highlighted in Appendix A1.1, which features a comparison of the laminar flame speed of all the fuels investigated in this study.

In Figure 5.16 and Figure 5.17, simulated HRR of gasoline are compared to measurement data at CR = 13. Again, the same model calibration parameters as for methanol were used here in their refined version. As shown in Figure 2.12, the *Lewis* or *Markstein* number of methanol and gasoline are significantly different. Nevertheless, the change in fuel could be predicted well, which again proves that a fuel influence on flame wrinkling does not need to be considered in the burn rate model. This is supported by the comparison of EGR and λ variation in Figure 5.16 and Figure 5.17, respectively: with an increasing EGR rate, the *Lewis* number does not change. In contrast, a significant change is ob-

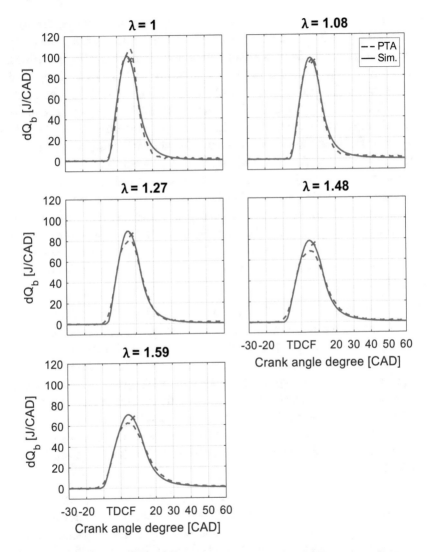

Figure 5.15: HRR change with λ for methyl formate (MeFo), engine C, 2500 rpm [72] (recalculated, edited)

served for a λ-variation of gasoline, theoretically causing a stronger decrease of s_T with λ (see Figure 2.12). Since both variations were predicted equally well, the approach to dismiss a fuel influence on flame wrinkling is verified. Figure 5.17 furthermore proves that the influence of EGR on s_L is covered well by the laminar flame speed model and the *Cai et al.* [22] reaction mechanism. Only at 19.7 % and 24.2 % EGR rate was the HRR overestimated in the simulation results. In comparison with the HRR change from 3.6 % to 16 % EGR rate, the strong HRR reduction in PTA when increasing the EGR rate from 16 % to only 19.7 % is surprising. This unexpected behavior could also be observed in the engine efficiency drop depicted in Figure 5.18. Furthermore, the almost instant increase in COV$_{IMEP}$ at a variation of both λ and especially EGR rate contradicts the typical exponential increase as was observed for methanol in Figure 5.13. This might indicate that the engine operation at 19.7 % and 24.2 % EGR rate was already influenced by CCV effects. However, no definite reason could be identified, since only averaged cylinder pressure traces were available.

The high prediction quality displayed in Figure 5.16 and Figure 5.17 furthermore highlights the capability of the model to properly react to engine configuration changes. Here, the compression ratio was reduced from 14.7 to 13. In Figure 5.14, a deformation of the heat release rate was observed and related to the piston geometry. In Figure 5.16 and Figure 5.17, this deformation is also visible in the measurement data, but is less prominent. Due to the changed CR, the theory of piston geometry influence seems plausible. As the λ-variation was measured at the knock limit (see MFB50 in Figure 5.18), the HRR of the AWC were additionally influenced by the HRR of auto-ignition of the knocking SWC.

Figure 2.8 depicts a theoretical comparison of lean combustion and increasing EGR rates on the basis of laminar flame speed changes. A similar investigation was made in [104]. It was concluded that, resulting from the lower s_L-reduction (and besides thermodynamic effects), lean combustion leads to a higher engine efficiency at a constant degree of mass dilution, compared to EGR. Furthermore, due to the higher flame speed and consequently lower COV$_{IMEP}$, higher degrees of dilution are possible. This investigation can now be validated using engine test bed data.

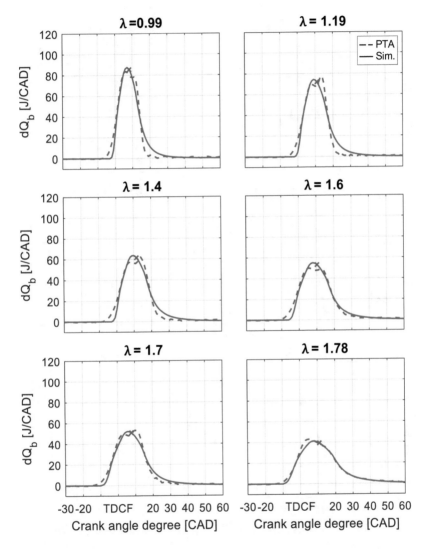

Figure 5.16: HRR change with λ, engine C, 2000 rpm, gasoline [72] (recalculated, edited)

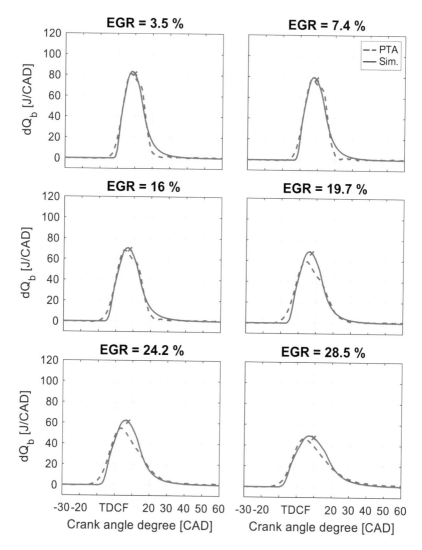

Figure 5.17: HRR change with EGR rate, engine C, 2000 rpm, gasoline [72]
(recalculated, edited)

Figure 5.18 shows the engine efficiency for increasing values of EGR rate and λ. The scaling of the x-axes corresponds to a comparable degree of dilution, related to mass. The increase in EGR rate or λ was stopped when COV_{IMEP} exceeded a fixed limit. From the measurement data, it cannot only be seen that the engine efficiency for lean conditions is higher (despite a later MFB50 due to knock), but also that higher degrees of dilution are possible when using lean combustion. In simulation, the engine efficiency was predicted with an error of maximum 1 %. On the one hand, this shows that the laminar flame speed can be used as a first indicator for such comparisons. On the other hand, it underlines that the burn rate model is capable of reacting properly to all sorts of boundary conditions changes.

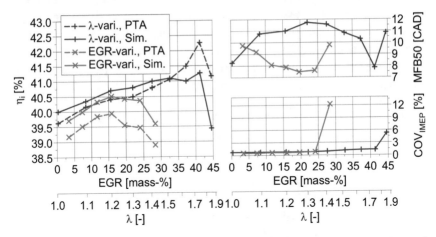

Figure 5.18: Comparison of EGR and λ: efficiency increase in simulation and PTA, engine C, 2000 rpm [72] (recalculated, edited)

In conclusion, the validation of the improved burn rate model using engines A, B and C proves that a variation of fuel can be predicted well by only using the laminar flame speed and flame thickness of the different fuels from reaction kinetics. The model calibration did not change significantly between the different engines despite a wide variation of boundary conditions such as fuel, λ, EGR, engine speed, turbulence level, engine load and compression ratio. Usually, changes in the calibration parameters can be related to a specific change in engine characteristics, like a reduction of the integral length scale due to a smaller combustion chamber or a reduced burn-up time due to lower CCV. The data of engine C not only illustrated that the burn rate model is still valid at high levels of turbulence, but also supported the theory that for engine A, the very high CCV could be the reason for the remaining prediction imperfections. Furthermore, it seems sufficient to only expand the models for the laminar flame speed and thickness in order to predict the influence of new fuels.

5.4 Engine D

The influence of DMC+ (65 vol–% dimethyl carbonate, 35 vol–% methyl formate) on the HRR was investigated on engine D. Its technical data and the operating conditions are given in Table 5.2. The engine measurement data were kindly provided by the TU Munich. The measurements and the model validation were performed in the course of the *NAMOSYN* project, funded by the German Federal Ministry of Education and Research.

Table 5.2: Engine D: technical data and operating conditions

Cylinder no.	1
Bore [mm]	82.51
Stroke [mm]	86.6
CR [-]	10.7
Engine speed [rpm]	2000
Engine load (IMEP [bar])	7.5
λ [-]	1
MFB50 [CAD]	7.5
Injection	DI

In order to match the HRR of gasoline, the burn rate model calibration parameters were set to $a_u = 2.6$, $a_\tau = 12$ and $a_{l_{int}} = 0.022$. The horizontal spark plug offset sp_{hor} was set to 8 mm. The influence of the local turbulence was not considered. The quality of the model calibration is underlined by the good agreement between measured and simulated HRR in Figure 5.19. Due to the lower laminar flame speed of DMC+ (see Appendix A1.1), the HRR is reduced in comparison to gasoline. This change could be predicted satisfactorily in simulation, without recalibrating the burn rate model. As for engines A, B and C, this again underlines the sufficiency of modeling the change in laminar flame speed and laminar flame thickness to predict the fuel influence on HRR, without accounting for an additional fuel influence on flame wrinkling.

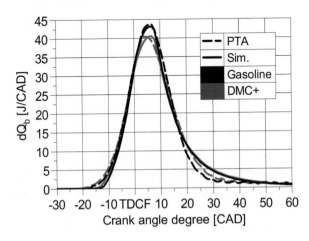

Figure 5.19: HRR change with DMC+, engine D

In addition to a fuel variation, a variation of injection timing was also performed on engine test bed. The variation showed a strong change of HRR with injection timing at otherwise constant operating conditions. This strong influence results from the lateral position of the fuel injector, which has a strong influence on the in-cylinder tumble motion, compared to a central injection. For DMC+, this influence is even more pronounced. Compared to gasoline, the lower heating value and lower stoichiometric air-fuel ratio of DMC+ necessitates a much higher injected fuel mass. Consequently, the injection duration and the ratio of fuel momentum to air momentum are increased. In addition, the higher evaporation enthalpy of DMC+ could also cause a difference in in-cylinder conditions. With that, the conditions before and after the intake valve are different, finally impacting the inflow of fresh air. More details were provided in [50], which presented a model to account for the fuel and injection influence on turbulence and mixture homogenization. Since the injection influence is reduced with fuel mass and thus engine load, the HRR change at

7.5 bar IMEP shown in Figure 5.19 should not be affected significantly by this phenomenon. Nevertheless, for an investigation of oxygenated fuels, this injection influence has to be considered, especially when using a lateral injector position, high engine loads or a variation of λ. For engine C, no significant influence of direct injection was observed, as the engine features a central injector position, which significantly reduces the impact on the tumble motion.

5.5 Engine E

For the investigation of water injection, data from the FVV project #1256 "Water Injection in SI Engines" were used in agreement with Dr. André Casal Kulzer (Porsche AG), the chairmen of the project. The PTA and the burn rate model calibration were performed by *Crönert* in [28]. The following investigation was already published by the author in [72]. Due to the updated laminar flame speed model, all simulations were repeated for this thesis. The technical data of the single-cylinder engine are given in Table 5.3. The burn rate model was calibrated to match the HRR at 0 % water injection (related to fuel mass). For this, the calibration parameters were set to $a_{l_{int}} = 0.015$, $a_u = 2.4$ and $a_\tau = 12$. The horizontal spark plug offset sp_{hor} was adjusted to 6 mm. The quality of the calibration is illustrated in Figure 5.20.

Table 5.3: Engine E: technical data and operating conditions

Bore [mm]	71.9
Stroke [mm]	82
CR [-]	11.65
Engine speed [rpm]	2500
Engine load (IMEP [bar])	15
MFB50 [CAD]	11

As discussed in Chapter 4.1, the direct influence of water injection on the laminar flame speed needs to be accounted for, in addition to the temperature reduction due to water evaporation. Figure 5.20 depicts the reduction of combustion speed caused by the injection of 30 % water, which leads to an increase in burn duration. This trend is covered by the simulation model, although the increase in burn duration is not strong enough. Usually, the water injection rate can be up to 100 %, which would affect the heat release rate even stronger due to a strong reduction of s_L. This underlines the necessity to cover the direct influence of water injection on the laminar flame speed. Furthermore, a reliable change in the laminar flame speed with temperature is necessary to properly predict the influence of water evaporation.

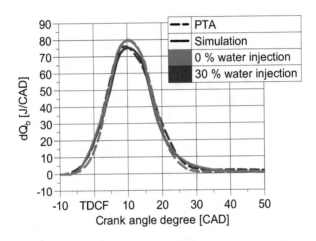

Figure 5.20: Water injection influence on HRR, engine E [72] (recalculated)

5.6 Engine F

The predictive ability of the improved burn rate model was investigated by the author in [73] for a large gas engine, which features a prechamber spark plug. Since the s_L-model was updated, the calibration parameters and results shown in the following are slightly different to [73]. The technical data of the engine is given in Table 5.4.

The engine was fueled with CNG and was operated at different λ, MFB50 and two different intake manifold air temperatures (IMAT). As illustrated in Figure 5.21, a good prediction of the ISFC was achieved by calibrating the model to the four highlighted operating points. In the calibration process, the

Table 5.4: Engine F: technical data and operating conditions

Fuel	Natural Gas (93 % CH$_4$)
Ignition	Prechamber spark plug
Bore / stroke [mm]	132 / 160
Engine speed @ 50 Hz [rpm]	1500
IMEP [bar]	19

homogeneous, isotropic k-ε turbulence model described in [62] was calibrated to roughly match the level and trend of the 3D CFD-based TKE published in [115]. Furthermore, the integral length was adapted. For passenger car engines, the integral length scale is modeled dependent on the cylinder volume via $l_{int} = (6/\pi \cdot V_{cyl})^{1/3}$ and scaled down to match not only the change in l_{int} with crank angle degree (CAD), but also the magnitude obtained from 3D CFD calculations (see Figure 4.19). However, 3D CFD calculations of Diesel-type combustion chambers (see [149]), similar to those of large gas engines, showed an almost constant l_{int} after TDCF, which contradicts the model approach for passenger cars. In consequence, l_{int} calculated in 3D CFD was imported as a function of CAD. At this stage, it is uncertain if this trend of l_{int} is dependent on engine geometry or the turbulence model used in 3D CFD. A possible explanation for a difference between passenger car (tumble concept) and Diesel-type combustion chambers (swirl concept) could be the different influence of piston movement on the flow characteristics. For tumble flow concepts, the vortex size is limited by the distance between piston and cylinder head, while the swirl flow restriction by the piston bowl size may be less severe.

The calibration parameter a_τ was set to 15. This way, the characteristic burn-up time value of about 1 ms was ensured. a_u had to be increased to a value of 5.3. This high value is needed to cover the increase in turbulence and flame surface area caused by the multiple flame cones exiting the prechamber spark plug. Again, the re-calibration of model parameters can be linked to a change in engine characteristics.

Simulated HRR are exemplarily compared to measurement data in Figure 5.22 for a variation of λ, while Figure 5.23 depicts this comparison for a variation of MFB50. Both variations could be modeled satisfactorily with a constant set

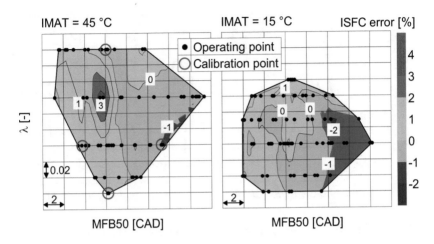

Figure 5.21: ISFC error of engine F [73] (recalculated)

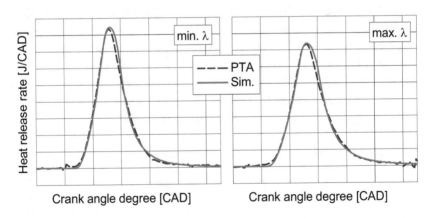

Figure 5.22: HRR change with λ, engine F [73] (recalculated)

of calibration parameters. At the start of combustion, a small error between measured and simulated HRR was observed. This is the result of the heat release in the prechamber, which was not considered in the simulation model.

Furthermore, the linking of the improved burn rate model to the CCV model was validated. The CCV model was calibrated using the same operating points

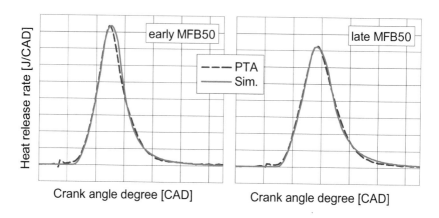

Figure 5.23: HRR change with MFB50, engine F [73] (recalculated)

as for the burn rate model. An automatic optimization was used to minimize the COV$_{IMEP}$ deviation by calibrating the two main model parameters χ_{ZS} and φ_{ZS} (see Chapter 4.7). The resulting parameters were comparable to those of passenger car engines. Figure 5.24 shows that the influences of changing λ, MFB50 and IMAT on COV$_{IMEP}$ were reproduced well for a constant set of model calibration parameters. This proves the validity of the changes made to the CCV model, which were described in Chapter 4.7.

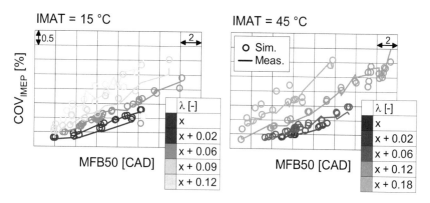

Figure 5.24: Sim. vs. meas. COV$_{IMEP}$, engine F [73] (recalculated)

5.7 Summary of Model Validation Conclusions

The validation of the improved burn rate model, using engines A to F, verified its high predictive quality when varying fuel, λ, EGR rate, water injection rate, engine speed, engine load, MFB50 and intake air temperature. This justifies the adapted approach to calculate the characteristic burn-up time, as presented in Chapter 4.4. Furthermore, the possibility to account for the local turbulent kinetic energy distribution proved to be useful.

In Chapter 3, a fuel influence on flame wrinkling was already discarded, based on a measurement data analysis. This conclusion was confirmed by the reliable prediction of changing fuels on engines A, B, C and D as well as the comparison of a λ and EGR variation on engine C. Consequently, it is sufficient to model the fuel influence on the laminar flame speed s_L and the laminar flame thickness δ_L. As illustrated in Appendix A1.1, the wide variation of fuels used for the model validation also covers a wide variation of laminar flame speed values. With that, the quality of the validation results promise the ability of the burn rate model to predict the influence of fuels not considered in this thesis by simply updating the models for s_L and δ_L.

The laminar flame speed model already showed a high quality in comparison with reaction kinetics calculation results (see Chapter 4.1.5). This quality was underlined by the very small change in IMEP error when importing the tabulated s_L and δ_L instead of using the model, as done on engines A and B. With that, the reaction mechanism is the last remaining uncertainty (compare Figure 2.7). The variation of engine load and speed on engine B highlighted the reliability of the *Cai et al.* mechanism concerning a change in the temperature and especially pressure. The validation of an increasing EGR rate on engine C and the variation of λ on engines A, C and F, as well as the increase in water injection rate on engine E, further indicated a proper coverage of boundary conditions influences when using the *Cai et al.* mechanism. Although a difference between measured and simulated HRR at rich conditions was observed for engine A, this error could not be entirely related to the reaction mechanism, as suggested in Figure 5.5. The fuel variations investigated on engines A, B and C underlined a proper change in s_L with the *Cai et al.* mechanism. Despite the use of a different mechanism for MeFo on engine C and DMC+ on engine D, the change in s_L still seemed to be plausible. It has to be stated that this evaluation of the reaction mechanism is dependent on the burn rate model structure. With a different s_T-model, for example, a different reaction mechanism could be more plausible. Nevertheless, the measured HRR changes with fuel and the results of the validation allow it to be said that the laminar flame speed given by the mechanisms published in [22] and [23] is plausible. This can be interpreted as an indirect determination of the laminar flame speed at engine relevant boundary conditions, which is not possible with direct laminar flame speed measurements (see Chapter 2.3.2).

While the influence of pressure and a changing laminar flame speed with λ on the turbulent flame speed model of *Peters* [117][119] were already validated in Chapter 4.2 and Figure 4.11 respectively, the validation using engine B confirmed the results. Moreover, the proper change in s_T with u' in the *Peters* model was proven by the validation using engine A and especially engine B, where u' was available from 3D CFD calculations.

The overall good validation results underline the validity of the entrainment approach in general (compare Chapter 2.1 and Chapter 2.2). Its mathematical interpretation, given in Chapter 4.4.2, is in accordance with changes made to the model calibration for engine C. Considering the different definitions of s_T, Chapter 2.4, where unburnt gas can be located behind the chosen reference flame front, the entrainment approach is reasonable and its phenomenological interpretation (see Chapter 4.4.1) as well as the s_T-model by *Peters* are in accordance with the entrainment theory.

5.8 Model Calibration Guideline

A guideline for the model calibration process was derived from the burn rate model validation. The calibration parameters are:

- a_u: scaling of the turbulent flame speed / entrainment velocity
- a_τ: scaling of the characteristic burn-up time τ_L
- $a_{l_{int}}$: scaling of the integral length scale l_{int}

- horizontal spark plug position sp_{hor}: calibration of the flame surface area calculation (if it cannot be calculated from 3D data)

- local u': a, m, $max(u'_{fac})$, $target$: changes u' with relative flame radius

a_u, a_τ and $a_{l_{int}}$ are the basic calibration parameters. sp_{hor} is only needed when the flame surface area cannot be calculated form 3D data. The influence of the local u' is not very significant, so it can be used for fine-tuning.

The model validation process showed that all changes made to these calibration parameters correspond to a changed engine characteristic. For engine F, a_u had to be increased significantly in order to account for the effects of the prechamber spark plug. For engine C, $a_{l_{int}}$ and thus l_{int} needed to be decreased, which represents the small combustion chamber resulting from the high CR of 14.7. This decrease in $a_{l_{int}}$ necessitated an increase in a_τ to maintain plausible values of the characteristic burn-up time. In Chapter 4.4, the linking of the burn-up time approach to an influence of CCV was mentioned. If the burn-up time is too low, the simulated HRR resembles more that of an SWC instead of an AWC. Nevertheless, it is reasonable to reduce a_τ for engines with very low CCV (e.g. engine C). As a guideline, typical variation ranges of the calibration parameters are given in Table 6. These ranges can be exceeded.

Table 5.5: Typical values of the burn rate model calibration parameters

Parameter	Standard	Range
a_u	3	2...4 4...7 (prechamber spark plug)
a_τ	12	8...16 for $l_{int} = 1$ mm, τ_L should be 0.2...1.5 ms
$a_{l_{int}}$	0.022	0.005...0.04 (so that $l_{int} = 0.2...2$ mm)
sp_{hor}	From 3D data or 8 mm	4...16 mm
Local u'	Off, or: a = 1 m = 0.25 $max(u'_{fac})$, $target = 1.2$	a: 0.7...5 @ m =0.25 m: 0...0.3 @ a = 1 $max(u'_{fac})$, $target = 1.1...1.4$

The calibration of the model to a new engine should always be performed using multiple measurement points in order to reduce the influence of measurement errors and uncertainties.

First, the model should be run with the standard calibration parameter values. Then, the minimum of the integral length scale and the characteristic burn-up time should be checked. These should be around 1 mm and 1 ms, respectively. These values correspond roughly to an engine with CR = 11, moderate turbulence level (burn durations of about 25 CAD) and COV_{IMEP} = 1-1.5 %. Depending on the differences between the new engine concept and this "base engine", the model parameters should be adapted according to the examples given above.

If the engine concept is more exotic, one possible way is to set l_{int} via $a_{l_{int}}$ to about 1 mm, and then perform a variation of a_u and a_τ at different values of sp_{hor}, if neither l_{int} from 3D CFD nor the flame surface from 3D data is available. By using an optimization algorithm, the best values of a_u and a_τ can be calculated easily for different values of sp_{hor}. The same procedure can be followed for a lower and a higher l_{int}, and the best result can be chosen by evaluating the IMEP error and the HRR shapes.

Fine-tuning can then be achieved by using the local u' correction and importing a different l_{int}-curve, see Figure 4.19. Here, it is also reasonable to consider the differences in the new combustion chamber to the one shown in Figure 4.16 to estimate the change in local turbulence.

Usually, the turbulence model described in [15] provides plausible values of u', similar to those of 3D CFD. When using the model published in [62], it is necessary to set a plausible turbulence level of about 2.6 m/s at 2000 rpm and TDCF. At best, the turbulence is available from 3D CFD calculations and can be used as a reference. However, it has to be kept in mind that the turbulent fluctuation velocity u' can differ by about ±20 % between roughly -20 to 60 CADaTDCF when using different 3D CFD codes for the same engine. These values are based on the experience of the author and are only intended as rough guidelines.

6 Engine Knock Investigation and Modeling

As validated in Chapter 5, the influence of a fuel variation on engine combustion characteristics can be predicted by only using fuel-specific models for the laminar flame speed and thickness. These characteristics represent, to a certain extent, the chemical effect of the fuel. To test a similar possibility for engine knock prediction, the fuel influence on engine knock is investigated in the following. Only a brief investigation is performed, since engine knock is beyond the main scope of this thesis.

A wide range of operating points taken from [38] (engine A) served as a measurement basis for the investigation. This range covers engine speeds of 1500 rpm to 3000 rpm, a boost pressure variation of 1 bar to 2 bar and a charge air temperature variation of 25 °C to 65 °C. Methane, methane/ethane mixtures, gasoline, toluene and ethanol were selected as fuels. In order to gain a better understanding of the physical processes influencing engine knock, possible effects of top land gas discharge on engine knock are evaluated first, before testing chemical effects by varying the fuel.

A comprehensive summary of the fundamentals of auto-ignition, engine knock and knock detection as well as numerous literature references were given by *Fandakov* in [42]. The most important aspects of those subjects are reproduced briefly in the following. Engine operation and optimization are limited by the phenomenon of engine knock. In-cylinder pressure oscillations are a key characteristic of this phenomenon, which can lead to engine damage. Engine knock is always preceded by auto-ignition, initiated by an oxidation of fuel without an external energy source initiating the process. The time until the event of auto-ignition, named ignition delay time, is dependent on boundary conditions such as temperature, pressure, mixture composition and the fuel itself. High temperatures or pressures, for example, reduce the ignition delay time. However, an auto-ignition does not necessarily result in a detonation, which causes the typical pressure oscillations. The consequence of an auto-ignition is dependent on the rate of energy release and the amount of total energy release, among other influences. A slow auto-ignition with little energy release can result in a regular deflagrative combustion, therefore acting like a spark plug. With the transition from deflagrative combustion to detonation being continuous, a limit needs to be defined to identify a single working cycle as a knocking one. This limit is linked to the selected knock detection method. A knock frequency limit is defined depending on the engine's endurance. The knock frequency describes the percentage of knocking single working cycles. If this limit is exceeded (for a longer period of time), engine damage is to be expected. Again, this limit is dependent on the knock detection method, since the number of single working cycles detected as knocking might vary. For the prediction of this knock limit in engine simulation, this dependency alone necessitates the calibration of the model to measurement data.

6.1 Investigation of Top Land Gas Discharge

Figure 6.1 (right) shows the difference (in CAD) between knock onset (KO) and peak pressure location (CA@p_{max}) for 1743 knocking SWC. The knock onset was determined by using the approach published in [39]. The comparison of KO and CA@p_{max} was done to investigate a possible influence of top land gas discharge on knock onset. A positive pressure gradient (which occurs before peak pressure) causes unburnt gas to be deposited in the top land volume. Due to the relatively cold cylinder wall and piston, its temperature is expected to be below the regular unburnt temperature. A negative pressure gradient (occurring after peak pressure) leads to a discharge of the top land gas. The top land gas could cool down the hot spot that would auto-ignite otherwise. In consequence, knock could either be suppressed or the knock onset could be delayed. However, Figure 6.1 shows that knock can occur after peak pressure, in this investigation in 253 SWC. Possible reasons are either the absence of a knock suppression effect by top land gas discharge or a knock occurrence before the cooling effect can take place, e.g. due to a certain mixing time. The absence of a knock suppression effect of top land gas discharge would support the theory of *Kleinschmidt* [123], where knock is assumed to occur in a large end gas pocket instead of an unburnt volume equally distributed at the circumference of the piston. In a large end gas pocket, the fraction of top land gas would be small, as most of the top land gas would be discharged into already burnt gas. In contrast, all the top land gas would mix with the whole unburnt mass in a circumferential unburnt volume. In that case, a cooling effect would be more likely. The schematic geometric relations for this case are illustrated in Figure 6.1 (left). A third possibility could be that the cooling effect is superimposed by other effects and thus not prominent.

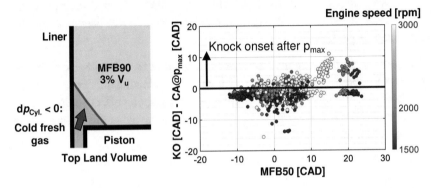

Figure 6.1: Left: schematic flame propagation at MFB90, right: relationship between peak pressure location CA@p_{max} and knock onset KO

If knock occurs before peak pressure is reached, the knock phenomenon itself causes the peak pressure. In this case, a local pressure maximum could be present before peak pressure, resulting in a negative pressure gradient between both maximums, theoretically allowing for the occurrence of top land gas discharge. This pressure characteristic was observed for only 20 SWC and hence does not distort the findings.

Another effect of top land gas discharge could be the reduction of knock intensity due to a partial cooling of the remaining unburnt gas. Partial cooling would reduce the amount of mass available for an (almost) simultaneous auto-ignition, consequently reducing the rate of energy release. An increase in time between peak pressure (start of top land gas discharge) and KO, allowing for a longer time for cooling effects to take place, might therefore decrease the knock intensity. With decreasing engine speed, the time for cooling effects at constant difference between KO and peak pressure would increase too, causing similar effects. Although higher knock intensities occurred in the measurement data at KO before peak pressure, no clear trend of knock intensity change with an increasing difference between KO and peak pressure location or engine speed could be observed. The identification of a trend is impeded by the overall significant scattering of the knock intensity and the relatively low number of operating points with KO after peak pressure. Furthermore, other influences like the unburnt mass fraction at knock onset have an influence on the knock intensity. With that, an influence of top land gas discharge on the knock intensity seems unlikely, but further investigations are needed to reliably evaluate this possibility.

The analysis of measurement data alone only allows a conclusion concerning the knock suppression effect and knock intensity reduction effect of discharged top land gas. But besides knock suppression, the cooling effect could lead to a delay in KO, rather than suppressing knock. To investigate this possibility, an auto-ignition model needs to be employed.

6.2 Expansion of an Auto-Ignition Model

For a first investigation of the fuel influence on engine knock and the possibility of delayed KO by top land gas discharge, the auto-ignition model for gasoline published by *Fandakov* in [42], [43] and [44] and the auto-ignition model for CNG published by *Urban et al.* in [137] were employed. Both models use mathematical approximations of ignition delay times from reaction kinetics calculations to cover the influence of different types of gasoline and different CNG compositions. Within their relatively small range of fuel variation, the respective approaches work well. In order to cover a wider range of fuels, an auto-ignition model was developed by using the expanded *Arrhenius* equation (eq. 6.1).

$$\tau = A \cdot \left(\frac{p}{100\,\mathrm{bar}}\right)^{\alpha} \lambda^{\beta} \cdot (1 - Y_{\mathrm{EGR}})^{\gamma} \cdot e^{\left(\frac{E_A}{R \cdot T_u}\right)} \qquad \text{eq. 6.1}$$

For some fuels, the non-exponential temperature dependency of their ignition delay time made the use of the *Weisser* approach [145] (eq. 6.2) necessary. The same model equations were also used by *Urban* [137].

$$\frac{1}{\tau} = \frac{1}{\tau_1 + \tau_2} + \frac{1}{\tau_3} \qquad \text{eq. 6.2}$$

The equations were calibrated to ignition delay times from reaction kinetics calculations using the mechanisms of *Cai et al.* [22] [23]. The resulting parameters are listed in Table 6.1 for toluene and ethanol and in Appendix A8.1 for methanol, methyl formate, DMC+ and H_2. If parameters are only listed for τ_1, the *Weisser* approach was not employed. Using these parameters, the model quality shown in Figure 6.2 for $\lambda = 1$ was achieved. At different values of λ or at higher rates of EGR or water injection, the error does not change significantly. A validation of the ignition delay time models for toluene and ethanol on the basis of engine measurements will be performed in the following subchapters. Due to a lack of measurement data, the models for all other fuels could only be validated with respect to reaction kinetics calculations. Nevertheless, the model parameters are given in Appendix A8.1 in order to cover the same range of fuels in both the burn rate model and the ignition delay time model. For methyl formate and DMC+, it is possible not to use the *Weisser* approach. As a consequence, the ignition delay time model error would increase by 5 to 10 percentage points.

Table 6.1: Ignition delay time model parameters of toluene and ethanol

τ_i [ms]	Parameter	Toluene	Ethanol
τ or τ_1	A [ms]	2.9298E-17	8.9817E-12
	α [-]	-1.0921	-0.7831
	β [-]	2.5115	0.5959
	γ [-]	-3.8440	-0.5980
	E_A/R [K]	32773	17910
τ_2	A [ms]	3.1057E-06	-
	α [-]	-0.9390	-
	β [-]	0.7200	-
	γ [-]	-1.8308	-
	E_A/R [K]	13037	-
τ_3	A [ms]	7.6516E-11	-
	α [-]	-0.6692	-
	β [-]	0.4473	-
	γ [-]	-1.8174	-
	E_A/R [K]	27528	-

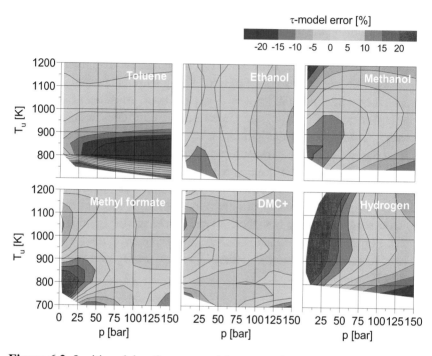

Figure 6.2: Ignition delay time τ: model vs. reaction kinetics calculations at $\lambda = 1$

To calculate the *Livengood-Wu* integral [94] (eq. 6.3), eq. 6.1 or eq. 6.2 was used. If the integral reaches a value of 1, auto-ignition occurs. This was verified by *Urban et al.* [137] by modeling combustion chamber processes using reaction kinetics calculations.

$$I_{K} = \int_{t=t_0}^{t=t_E} \frac{1}{\tau} \cdot dt \qquad \text{eq. 6.3}$$

6.3 Knock Onset Prediction: Single Working Cycles

By comparing the crank angle degree where $I_K = 1$ (see eq. 6.3) with the knock onset determined in measurement, the quality of the reaction kinetics-based auto-ignition model was evaluated for SWC. This comparison is displayed in Figure 6.3. In the model, a temperature offset of 10 % was used, representing a hot spot. The model was applied to the temperature and pressure trace of SWC from PTA of engine A. For all fuels, the knock onset was predicted well. The offset of the fuel "AvGas100LL" is a result of its uncertain RON and MON and the utilization of a toluene reference fuel to cover the influence of the knock propensity reduction by the addition of tetraethyl lead in the fuel.

The good agreement between measured KO and the KO from reaction kinetics (CAD where $I_K = 1$), illustrated in Figure 6.3, proves that the fuel influence on engine knock is a result of the different chemistry, which is represented in simulation by a model of the ignition delay time from reaction kinetics calculations. The scattering of the KO of the SWC can have multiple reasons. For example, model uncertainties in terms of ignition delay time model or pressure trace analysis (e.g. mixture composition) of SWC could lead to these fluctuations. Furthermore, changing characteristics of the auto-ignition spot due to CCV can cause a difference between measurement and simulation. According to *Kleinschmidt* [123], knock is most likely to occur in large end gas pockets. The characteristics of these pockets, e.g. temperature, mixture composition, size etc., are strongly dependent on CCV in terms of flame propagation and mixture inhomogeneities. Up to now, only the temperature of the unburnt zone (see Figure 2.1) and the average mixture composition serve as inputs for the auto-ignition model and no local effects or CCV influences are accounted for, but these could improve the model quality.

The SWC with KO after peak pressure show the same scattering as with KO before peak pressure in Figure 6.3. If a cooling effect would delay the KO, the simulated KO should be too early because of the higher knock tendency at higher temperatures. This either means, again, that the time for the cooling effect to take place is too short, or that there is no influence of top land gas discharge and knock occurs in a large end gas pocket instead of a circumferential volume, plotted schematically in Figure 6.1 (left).

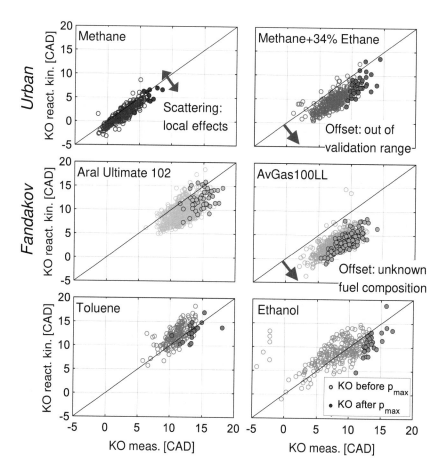

Figure 6.3: Predicted vs. measured knock onset (KO) of single working cylces [72] (recalculated, expanded)

6.4 Knock Onset Prediction: Averaged Working Cycles

For the investigation of averaged working cycles (AWC), the measurement database was reduced in order to obtain constant engine operation conditions. Those were an IMEP of 16 bar and a boost pressure of about 1.5 bar. The charge air temperature needed to be reduced from 45 °C for gaseous fuels to 25 °C for liquid fuels due to the high knock propensity. For this reason, gasoline also had to be excluded from this investigation.

For the investigation of AWC, a change in the evaluation of the *Livengood-Wu* integral I_K is necessary. For single working cycles (SWC), the crank angle degree at $I_K = 1$ is evaluated. For an AWC, a limit is needed up to which I_K is calculated. If $I_K < 1$ at this evaluation limit, knock did not occur in the AWC. In the measurement data, the knock onset of the AWC was found to be always before peak pressure (in contrast to the KO of SWC), as depicted in Appendix A8.2. In consequence, the upper evaluation limit of I_K was set to the crank angle at peak pressure. In Figure 6.4, the change in I_K in simulation is compared to the increase in measured knock frequency for changing MFB50 at different fuels. For earlier MFB50, both the knock frequency and the I_K-value increased in measurement and simulation, respectively. When fixing the knock limit to 5 % knock frequency, $I_K = 1$ was reached for methane. In simulation, this can be used to control the MFB50 at knock limit. If $I_K > 1$, the MFB50 is set to a higher value. For $I_K < 1$, the MFB50 can be set earlier. When changing the fuel to methane + 34 vol − % ethane, the MFB50 at knock limit was predicted with an error of only 1 CAD. Ethanol and toluene were predicted as too knock resistant, as I_K never reached the value of one, although the knock frequency was similar for all fuels. Furthermore, due to the evaluation of I_K at peak pressure, the knock tendency of ethanol should have been predicted as too high, since peak pressure was reached much later compared to knock onset in the measurements (see Appendix A8.2). This would result in a longer time period where the I_K-integral is calculated. With that, higher I_K-values would be reached, which corresponds to a higher predicted knock frequency. As shown in Figure 6.4, the error for toluene and ethanol decreased with increasing engine speed. Considering the good prediction quality of KO in SWC for both fuels at different engine speeds, the error in the prediction of AWC is surprising.

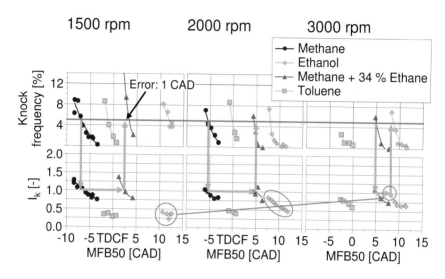

Figure 6.4: Prediction of the knock limit using averaged working cycles, engine A [72] (edited)

The impression arises that some information is lost when using AWC instead of SWC. To identify the type of information, the knock onsets of the AWC were investigated in simulation for all operating points. For this, the AWC were calculated using all SWC. The simulated knock onsets of these AWC were compared to those of the AWC where only not knocking SWC were considered for averaging. The knock onsets with and without knocking SWC did not show any difference, as highlighted in Appendix A8.3. Consequently, no information of the knock frequency is included in the pressure trace of an AWC. In conclusion, averaging the SWC results in the CCV level information being lost.

Figure 6.5 illustrates the COV_{IMEP} and COV_{pmax} of all the fuels for all engine speeds. While both COV of methane and methane + 34 vol−% ethane were similar, a strong change in COV_{IMEP} and especially COV_{pmax} for toluene and especially ethanol with engine speed was observed. This can explain the underestimation of the knock tendency of ethanol and toluene in simulation. The higher CCV result in a stronger change in end gas pocket characteristics and lead to some SWC having a higher pressure and, with that, a higher temperature. Those effects cause these SWC to knock, but no information about the CCV level is included in the AWC. Due to the lack of this information, ethanol and toluene were considered overly knock resistant in simulation. At the highest engine speed, the CCV of all fuels were similar. Figure 6.4 shows that all fuels were properly predicted at that engine speed, even toluene, as its knock frequency was below 5 %, resembling an I_K <1. A simulated knock limit is thus linked to a certain level of CCV. The sensitivity of this link is even higher at the early MFB50 illustrated in Figure 6.4. At early MFB50, a slight change in MFB50 does not significantly change the IMEP, but leads to a strong change in peak pressure. In conclusion, a change in fuel can be predicted in SWC and AWC by using ignition delay times from reaction kinetics. However, for AWC, the simulation of MFB50 at knock limit is linked to a certain level of CCV.

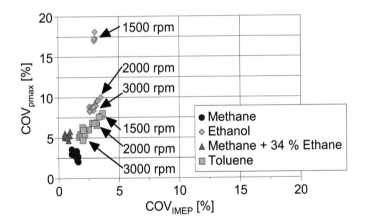

Figure 6.5: COV_{IMEP} and COV_{pmax}, engine A [72] (edited)

7 Conclusion and Outlook

In this thesis, a quasi-dimensional spark-ignition burn rate model for predicting the effects of changing fuel, air-fuel ratio λ, exhaust gas recirculation (EGR) and water injection was developed and validated for a wide range of boundary conditions. In the validation process, a high predictive ability was proven, especially concerning the prediction of fuel influence on engine performance. The investigated fuels were methane, CNG substitutes, methanol, ethanol, gasoline, hydrogen, methyl formate and DMC+ (65 vol–% dimethyl carbonate, 35 vol–% methyl formate). Based on the model calibration for one fuel at one λ, the change of heat release rate (HRR) with a variation of fuel and λ could be predicted, without recalibration. This ability was achieved by modeling the laminar flame speed and laminar flame thickness of the different fuels on the basis of reaction kinetics calculations. A systematic measurement data analysis proved this approach to cover the fuel influence in engine simulation to be sufficient, since a fuel influence on flame wrinkling could not be observed. Additionally, the baseline turbulent flame speed model was evaluated in terms of its reaction to fundamental changes in boundary conditions taken from the literature. Due to its poor performance, it was replaced by a more sophisticated model. Furthermore, the approach to calculate the characteristic burn-up time was improved and interpreted phenomenologically as well as mathematically. This interpretation clarified the influence of single working cycles (SWC) on the HRR of an averaged working cycle (AWC), which resembles an influence of cycle-to-cycle variations (CCV). Due to the importance of CCV, the burn rate model was linked to an existing CCV model. By implementing a model to account for the local distribution of turbulence, the burn rate model was improved further. In the framework of the improved burn rate model, the change in laminar flame speed with boundary conditions, as obtained from reaction kinetics calculations, proved to be plausible.

Furthermore, a brief investigation of engine knock was performed to test the method of exchanging the chemical fuel influence in the knock model, which, to a certain extent, corresponds to the exchange in laminar flame speed and thickness in the burn rate model. For this, an existing engine knock model

© The Author(s), under exclusive license to
Springer Fachmedien Wiesbaden GmbH, part of Springer Nature 2021
S. Hann, *A Quasi-Dimensional SI Burn Rate Model for Carbon-Neutral Fuels*,
Wissenschaftliche Reihe Fahrzeugtechnik Universität Stuttgart,

was expanded to also cover toluene and ethanol by expanding its submodel for ignition delay times, based on reaction kinetics calculations. The model was validated using engine measurements of different fuels from [38]. For a comparable level of CCV, the predictive ability of the model proved to be reliable. Based on the expanded knock model, the influence of top land gas discharge was investigated using single working cycles. Since neither knock suppression nor a delayed knock onset could be observed, an influence of top land gas discharge on engine knock was considered irrelevant. It was concluded that knock is therefore most likely to occur in a large end gas pocket.

However, a possible influence of top land gas discharge on knock intensity could not be ruled out with certainty. A more detailed investigation could be performed in future work by combining LES calculations with engine measurements specifically designed to isolate this influence. Furthermore, the influence of water injection is considered in the burn rate model for all fuels, but only in the auto-ignition model of gasoline. With this influence being of interest for the development of future engine concepts, it should be added to the model and validated for all fuels.

Future work could also focus on the slight underestimation of heat release rates at rich conditions, which was identified in the validation using data of engine A. As this engine showed signs of a significant wall film build-up, resulting in an oscillating air-fuel ratio measurement, the operating conditions were too uncertain to find a definite reason for the model error. New engine measurements at higher intake air temperatures, consequently reducing the wall film, could lead to new insights.

Another interesting subject is the further improvement of the model to account for the local turbulent kinetic energy (TKE). The turbulence model published in [15] is based on the *Taylor-Green* eddy [134] and already calculates a TKE distribution for a simplified combustion chamber geometry. This distribution could be used as a reference for the imposed correction factor, which was modeled using a 3D CFD-based, mathematical approach in this thesis. In a next step, more detailed combustion chamber geometries could be implemented in the *Taylor-Green* eddy calculation, possibly allowing to directly calculate a reliable distribution of the TKE. Another approach could be to calculate a target TKE distribution by using an iteration process to match the (well-calibrated) simulated heat release rate to the measured one. This target distribution could then be checked for plausibility and might allow a perfect match of simulated and measured heat release rates.

In Chapter 4.4.2, the necessity for the general entrainment approach was linked to the influence of CCV by the interpretation of the characteristic burn-up time. This influence also translates to single working cycles impacting the shape of the averaged working cycle heat release rate. In a first step, it would be interesting to evaluate the benefit of a CCV-dependent model calibration with the aim of matching the heat release rate despite a change in CCV. In a next step, it might be possible to calculate the heat release rate of an AWC without using the entrainment approach. Instead, several single working cycles could be simulated and averaged. As suggested in Figure 4.14, even a simple averaging method already gives promising results. To achieve results of sufficient quality, local effects that also change with CCV might have to be considered. In the class of 0D/1D models, additional submodels to cover the change of these effects with operating conditions are not available yet and need to be developed.

Although the need for a distinct model for the early flame development phase is questionable when considering the influence of single working cycles on the heat release rate of an averaged working cycle highlighted in Chapter 4.4.2, the model validation shows a slight heat release rate error at low MFB. If future investigations allow the influence of single working cycles to be discarded as a reason, several models from the literature could be evaluated. In this context, the strong flame curvature and its presumably overall positive value at low mass fractions burnt might make it necessary to account for an additional fuel influence.

Due to a lack of measurement data, the burn rate model could not be validated for pure hydrogen or mixtures of gasoline and ethanol. Especially the validation of hydrogen is of great interest for future work, as the laminar flame speed is about a factor of 10 above that of hydrocarbons. Furthermore, hydrogen is the only fuel considered in this thesis with a *Lewis* number below one, consequently exhibiting an opposite trend of theoretical fuel influence on flame wrinkling compared to higher hydrocarbons. This would allow to re-evaluate the approach to model the fuel influence solely by a change in laminar flame speed and thickness. The flame thickness model for hydrogen should be re-evaluated too, since the thickness definition used for hydrocarbons could not be applied directly to hydrogen, causing a possible source of error in the model. Nevertheless, considering the wide range of fuels already used for model validation, which not only cover a considerable range of laminar flame speeds, but also a significant variation of the *Lewis* number, reliable simulation results for hydrogen are likely. This assumption is underlined by the model validation results for mixtures of methane with up to 30 mol−% hydrogen, as illustrated in Figure 5.8. However, additional effects can have an influence on the burn rate. Direction injection, especially of gaseous fuels, can cause inhomogeneities, which influence the local flame propagation and, with that, possibly also the burn rate. Depending on the injector targeting, direct injection might also influence the in-cylinder turbulence level, as already mentioned in Chapter 5.4 for a liquid, oxygenated fuel. A research project is already planned in order to investigate this influence for hydrogen, as well as the formation of inhomogeneities.

Bibliography

[1] P. G. Aleiferis, J. Serras-Pereira, and D. Richardson. Characterisation of flame development with ethanol, butanol, iso-octane, gasoline and methane in a direct-injection spark-ignition engine. *Fuel*, 109:256–278, 2013.

[2] P. G. Aleiferis, A. Taylor, K. Ishii, and Y. Urata. The nature of early flame development in a lean-burn stratified-charge spark-ignition engine. *Combustion and Flame*, 136(3):283–302, 2004.

[3] American Society for Testing and Materials International. ASTM D5798: Standard Specification for Ethanol Fuel Blends for Flexible-Fuel Automotive Spark-Ignition Engines, 2017.

[4] R. Amirante, E. Distaso, P. Tamburrano, and R. D. Reitz. Laminar flame speed correlations for methane, ethane, propane and their mixtures, and natural gas and gasoline for spark-ignition engine simulations. *International Journal of Engine Research*, 18(9):951–970, 2017.

[5] J. E. Anderson, D. M. DiCicco, J. M. Ginder, U. Kramer, T. G. Leone, H. E. Raney-Pablo, and T. J. Wallington. High octane number ethanol–gasoline blends: Quantifying the potential benefits in the United States. *Fuel*, 97:585–594, 2012.

[6] T. Badawy, J. Williamson, and H. Xu. Laminar burning characteristics of ethyl propionate, ethyl butyrate, ethyl acetate, gasoline and ethanol fuels. *Fuel*, 183:627–640, 2016.

[7] M. Baloo, B. M. Dariani, M. Akhlaghi, and M. AghaMirsalim. Effects of pressure and temperature on laminar burning velocity and flame instability of iso-octane/methane fuel blend. *Fuel*, 170:235–244, 2016.

[8] M. Bargende. *Ein Gleichungsansatz zur Berechnung der instationären Wandwärmeübergänge im Hochdruckteil von Ottomotoren*. PhD thesis, Technische Hochschule Darmstadt, 1991.

© The Editor(s) (if applicable) and The Author(s), under exclusive license to
Springer Fachmedien Wiesbaden GmbH, part of Springer Nature 2021
S. Hann, *A Quasi-Dimensional SI Burn Rate Model for Carbon-Neutral Fuels*,
Wissenschaftliche Reihe Fahrzeugtechnik Universität Stuttgart,

[9] M. Bargende, H.-C. Reuss, and J. Wiedemann, editors. *16. Internationales Stuttgarter Symposium Automobil- und Motorentechnik.* Proceedings. Springer Vieweg, Wiesbaden, 2016.

[10] J. K. Bechtold and M. Matalon. The dependence of the Markstein length on stoichiometry. *Combustion and Flame*, 127(1-2):1906–1913, 2001.

[11] J. B. Bell, R. K. Cheng, M. S. Day, and I. G. Shepherd. Numerical simulation of Lewis number effects on lean premixed turbulent flames. *Proceedings of the Combustion Institute*, 31(1):1309–1317, 2007.

[12] F. Bergk, K. Biemann, C. Heidt, W. Knörr, U. Lambrecht, T. Schmidt, L. Ickert, M. Schmied, P. Schmidt, and W. Weindorf. Climate Change Mitigation in Transport until 2050 (German): UBA-FB 002355.

[13] G. Blanquart. Effects of spin contamination on estimating bond dissociation energies of polycyclic aromatic hydrocarbons. *International Journal of Quantum Chemistry*, 115(12):796–801, 2015.

[14] N. Blizard and J. C. Keck. Experimental and Theoretical Investigation of Turbulent Burning Model for Internal Combustion Engines. *SAE Technical Paper*, 740191, 1974.

[15] C. Bossung, M. Bargende, and O. Dingel. A quasi-dimensional charge motion and turbulence model for engine process calculations. *Proceedings of the 15th Stuttgart International Symposium*, 2015.

[16] S. Bougrine, S. Richard, A. Nicolle, and D. Veynante. Numerical study of laminar flame properties of diluted methane-hydrogen-air flames at high pressure and temperature using detailed chemistry. *International Journal of Hydrogen Energy*, 36(18):12035–12047, 2011.

[17] F. Bozza, A. Gimelli, S. S. Merola, and B. M. Vaglieco. Validation of a Fractal Combustion Model through Flame Imaging. *SAE Technical Paper*, 2005-01-1120, 2005.

[18] D. Bradley, M. Z. Haq, R. A. Hicks, T. Kitagawa, M. Lawes, C. Sheppard, and R. Woolley. Turbulent burning velocity, burned gas distribution, and associated flame surface definition. *Combustion and Flame*, 133(4):415–430, 2003.

[19] D. Bradley, M. Lawes, and M. S. Mansour. Correlation of turbulent burning velocities of ethanol–air, measured in a fan-stirred bomb up to 1.2MPa. *Combustion and Flame*, 158(1):123–138, 2011.

[20] P. Brequigny, F. Halter, C. Mounaïm-Rousselle, and T. Dubois. Fuel performances in Spark-Ignition (SI) engines: Impact of flame stretch. *Combustion and Flame*, 166:98–112, 2016.

[21] U. Burke, W. K. Metcalfe, S. M. Burke, K. A. Heufer, P. Dagaut, and H. J. Curran. A detailed chemical kinetic modeling, ignition delay time and jet-stirred reactor study of methanol oxidation. *Combustion and Flame*, 165:125–136, 2016.

[22] L. Cai and H. Pitsch. Optimized chemical mechanism for combustion of gasoline surrogate fuels. *Combustion and Flame*, 162(5):1623–1637, 2015.

[23] L. Cai, F. Vom Lehn, K. A. Heufer, and H. Pitsch. Reaction Mechanism for DMC+: preliminary result of the project "NAMOSYN", 2020.

[24] A. Cairns, P. Stansfield, N. Fraser, H. Blaxill, M. Gold, J. Rogerson, and C. Goodfellow. A Study of Gasoline-Alcohol Blended Fuels in an Advanced Turbocharged DISI Engine. *SAE International Journal of Fuels and Lubricants*, 2(1):41–57, 2009.

[25] S. Chaudhuri, V. Akkerman, and C. K. Law. Spectral formulation of turbulent flame speed with consideration of hydrodynamic instability. *Physical review. E, Statistical, nonlinear, and soft matter physics*, 84(2 Pt 2):026322, 2011.

[26] S. Chaudhuri, F. Wu, and C. K. Law. Scaling of turbulent flame speed for expanding flames with Markstein diffusion considerations. *Physical review. E, Statistical, nonlinear, and soft matter physics*, 88(3):033005, 2013.

[27] S. Crönert. *Modeling of Future Fuels for Quasi-Dimensional Engine Simulation*. Master's thesis, University of Stuttgart, Stuttgart, 2019.

[28] S. Crönert. *Reaction Kinetic Study and Modeling on the Impact of Water Injection on the Laminar Flame Speed of Hydrocarbon Compounds.* Student research project, University of Stuttgart, Stuttgart, 2019.

[29] G. Damkoehler. The Effect of Turbulence on the Flame Velocity in Gas Mixtures: National Advisory Committee for Aeronautics 1112.

[30] G. Darrieus. Propagation d'un front de flamme. *La Technique Moderne*, 30:18, 1938.

[31] S. Demesoukas, C. Caillol, P. Higelin, and A. Boiarciuc. Zero-Dimensional Spark Ignition Combustion Modeling - A Comparison of Different Approaches. *SAE Technical Paper*, 2013-24-0022, 2013.

[32] Deutsches Institut für Normung. EN 228:2012+A1:2017: Automotive fuels - Unleaded petrol - Requirements and test methods; German version.

[33] V. Di Sarli and A. Di Benedetto. Laminar burning velocity of hydrogen–methane/air premixed flames. *International Journal of Hydrogen Energy*, 32(5):637–646, 2007.

[34] P. Dirrenberger, P. A. Glaude, R. Bounaceur, H. Le Gall, A. P. da Cruz, A. A. Konnov, and F. Battin-Leclerc. Laminar burning velocity of gasolines with addition of ethanol. *Fuel*, 115:162–169, 2014.

[35] P. Dirrenberger, H. Le Gall, R. Bounaceur, O. Herbinet, P.-A. Glaude, A. Konnov, and F. Battin-Leclerc. Measurements of Laminar Flame Velocity for Components of Natural Gas. *Energy & Fuels*, 25(9):3875–3884, 2011.

[36] B. Z. Dlugogorski, R. K. Hichens, E. M. Kennedy, and J. W. Bozzelli. Propagation of Laminar Flames in Wet Premixed Natural Gas-Air Mixtures. *Process Safety and Environmental Protection*, 76(2):81–89, 1998.

[37] J. Driscoll. Turbulent premixed combustion: Flamelet structure and its effect on turbulent burning velocities. *Progress in Energy and Combustion Science*, 34(1):91–134, 2008.

[38] M. Eberbach, M. Bargende, H.-J. Berner, F. Altenschmidt, and M. Schenk. Methane–Based Fuels and High–Octane Liquid Fuels during Knocking Combustion. *Proceedings of the 13th Conference on Gaseous-Fuel Powered Vehicles*, 2019.

[39] C. Elmqvist, F. Lindström, H.-E. Ångström, B. Grandin, and G. Kalghatgi. Optimizing Engine Concepts by Using a Simple Model for Knock Prediction. *SAE Technical Paper*, 2003-01-3123, 2013.

[40] Eoin M. Burke, Felix Güthe, and Rory F. D. Monaghan. A comparison of turbulent flame speed correlations for hydrocarbon fuels at elevated pressures. *Proceedings of ASME Turbo Expo 2016*, GT2016-57804, 2016.

[41] J. Ewald. *A level set based flamelet model for the prediction of combustion in homogeneous charge and direct injection spark ignition engines*. Ph.D. thesis, RWTH Aachen, 2006.

[42] A. Fandakov. *A Phenomenological Knock Model for the Development of Future Engine Concepts*. Wissenschaftliche Reihe Fahrzeugtechnik Universität Stuttgart. Springer Fachmedien Wiesbaden, Wiesbaden, 2019.

[43] A. Fandakov, M. Grill, M. Bargende, and A. C. Kulzer. Two-Stage Ignition Occurrence in the End Gas and Modeling Its Influence on Engine Knock. *SAE International Journal of Engines*, 10(4):2109–2128, 2017.

[44] A. Fandakov, M. Grill, M. Bargende, and A. C. Kulzer. A Two-Stage Knock Model for the Development of Future SI Engine Concepts. *SAE Technical Paper*, 2018-01-0855, 2018.

[45] J. T. Farrell, R. J. Johnston, and I. P. Androulakis. Molecular Structure Effects On Laminar Burning Velocities At Elevated Temperature And Pressure. *SAE Technical Paper*, 2004-01-2936, 2004.

[46] S. Fasse, S. Fritsch, Q. Yang, M. Grill, and M. Bargende. Quasi-dimensional Modeling of Complex Charge Motion and Fuel Injection Influence on SI Engine Turbulence. *Proceedings of the 17th Conference "The Working Process of the Internal Combustion Engine", Graz*, pages 461–472, 2019.

[47] A. Frassoldati, A. Cuoci, T. Faravelli, and E. Ranzi. Kinetic Modeling of the Oxidation of Ethanol and Gasoline Surrogate Mixtures. *Combustion Science and Technology*, 182(4-6):653–667, 2010.

[48] L. E. Freeman, R. J. Roby, and G. K. Chui. Performance and Emissions of Non–Petroleum Fuels in a Direct–Injection Stratified Charge SI Engine. *SAE Technical Paper*, 821198, 1982.

[49] M. Frenklach, H. Wang, M. Goldenberg, G. Smith, D. Golden, C. Bowman, R. Hanson, W. Gardiner, and V. Lissianski. GRI-Mech—An Optimized Detailed Chemical Reaction Mechanism for Methane Combustion, 2000. http://combustion.berkeley.edu/gri-mech/, 30.12.2020.

[50] S. Fritsch, M. Grill, M. Bargende, and O. Dingel. A Phenomenological Homogenization Model Considering Direct Fuel Injection and EGR for SI Engines. *SAE Technical Paper*, 2020-01-0576, 2020.

[51] B. Galmiche, F. Halter, and F. Foucher. Effects of high pressure, high temperature and dilution on laminar burning velocities and Markstein lengths of iso-octane/air mixtures. *Combustion and Flame*, 159(11):3286–3299, 2012.

[52] Gamma Technologies LLC. GT-Power. www.gtisoft.com, 08.06.2020.

[53] Gecko Instruments GmbH. Gasanalyse Erdgas Holland L. https://www.gecko-instruments.de/media/Erdgaszusammensetzung/Erdgas_Holland_L.pdf, 20.04.2020.

[54] Gecko Instruments GmbH. Gasanalyse Erdgas Nordsee H. https://www.gecko-instruments.de/media/Erdgaszusammensetzung/Erdgas_Nordsee_H.pdf, 20.04.2020.

[55] Gecko Instruments GmbH. Gasanalyse Erdgas Russland H. https://www.gecko-instruments.de/media/Erdgaszusammensetzung/Erdgas_Russland_H.pdf, 20.04.2020.

[56] Gecko Instruments GmbH. Gasanalyse Erdgas Weser Ems L. https://www.gecko-instruments.de/media/Erdgaszusammensetzung/Erdgas_Weser_Ems_L.pdf, 20.04.2020.

[57] P. J. Goix and I. G. Shepherd. Lewis Number Effects on Turbulent Premixed Flame Structure. *Combustion Science and Technology*, 91(4-6):191–206, 1993.

[58] D. G. Goodwin, R. L. Speth, H. K. Moffat, and B. W. Weber. Cantera: An Object-oriented Software Toolkit for Chemical Kinetics, Thermodynamics, and Transport Processes, 2018.

[59] J. Göttgens, F. Mauss, and N. Peters. Analytic approximations of burning velocities and flame thicknesses of lean hydrogen, methane, ethylene, ethane, acetylene, and propane flames. *Symposium (International) on Combustion*, 24(1):129–135, 1992.

[60] M. Grill. *Objektorientierte Prozessrechnung von Verbrennungsmotoren*. Ph.D. thesis, University of Stuttgart, 2006.

[61] M. Grill and M. Bargende. The Development of an Highly Modular Designed Zero-Dimensional Engine Process Calculation Code. *SAE International Journal of Engines*, 3(1):1–11, 2010.

[62] M. Grill, T. Billinger, and M. Bargende. Quasi-Dimensional Modeling of Spark Ignition Engine Combustion with Variable Valve Train. *SAE Technical Paper*, 2006-01-1107, 2006.

[63] M. Grill, M. Chiodi, H.-J. Berner, and M. Bargende. Calculating the Thermodynamic Properties of Burnt Gas and Vapor Fuel for User-Defined Fuels. *MTZ worldwide*, (68):30–35, 2007.

[64] M. Grill, D. Rether, and M.-T. Keskin. FKFS UserCylinder v2.6.3, www.usercylinder.com, 30.12.2020.

[65] M. Grill, A. Schmid, M. Chiodi, H.-J. Berner, and M. Bargende. Calculating the Properties of User-Defined Working Fluids for Real Working-Process Simulations. *SAE Technical Paper*, 2007-01-0936, 2007.

[66] G. R. Groot. *Modelling of propagating spherical and cylindrical premixed flames*. Ph.D. thesis, Eindhoven University of Technology, 2003.

[67] M. Gross, A. Mazacioglu, J. Kern, and V. Sick. Infrared Borescopic Analysis of Ignition and Combustion Variability in a Heavy-Duty Natural-Gas Engine. *SAE Technical Paper*, 2018-01-0632, 2018.

[68] Ö. L. Gülder. Correlations of Laminar Combustion Data for Alternative SI Engine Fuels. *SAE Technical Paper*, 841000, 1984.

[69] W. Han and Z. Chen. Effects of Soret diffusion on spherical flame initiation and propagation. *International Journal of Heat and Mass Transfer*, 82:309–315, 2015.

[70] S. Hann. *Reaktionskinetische Bestimmung laminarer Flammengeschwindigkeiten von binären, methanbasierten CNG-Substituten*. Master's thesis, HTWG Konstanz, 2016.

[71] S. Hann, M. Grill, and M. Bargende. Reaction Kinetics Calculations and Modeling of the Laminar Flame Speeds of Gasoline Fuels. *SAE Technical Paper*, 2018-01-0857, 2018.

[72] S. Hann, M. Grill, and M. Bargende. A Quasi-Dimensional SI Combustion Model for Predicting the Effects of Changing Fuel, Air-Fuel-Ratio, EGR and Water Injection. *SAE Technical Paper*, 2020-01-0574, 2020.

[73] S. Hann, M. Grill, M. Bargende, M. Veltman, and S. Palaveev. Predicting the Influence of Charge Air Temperature Reduction on Engine Efficiency, CCV and NOx-Emissions of a Large Gas Engine using a SI Burn Rate Model. *SAE Technical Paper*, 2020-01-0575, 2020.

[74] S. Hann, L. Urban, M. Grill, and M. Bargende. Influence of Binary CNG Substitute Composition on the Prediction of Burn Rate, Engine Knock and Cycle-to-Cycle Variations. *SAE International Journal of Engines*, 10(2):501–511, 2017.

[75] J. B. Heywood. *Internal Combustion Engine Fundamentals, Second Edition*. The McGraw-Hill Companies, Inc, 1988.

[76] P. G. Hill and J. Hung. Laminar Burning Velocities of Stoichiometric Mixtures of Methane with Propane and Ethane Additives. *Combustion Science and Technology*, 60(1-3):7–30, 1988.

[77] I. Hunwartzen. Modification of CFR Test Engine Unit to Determine Octane Numbers of Pure Alcohols and Gasoline-Alcohol Blends. *SAE Technical Paper*, 820002, 1982.

[78] J. Hustad and O. Sonju. Experimental studies of lower flammability limits of gases and mixtures of gases at elevated temperatures. *Combustion and Flame*, 71(3):283–294, 1988.

[79] J. J. Hyvönen. *Experimentelle und numerische Untersuchung magerer Methan-Hochdruckverbrennung unter Mikrogravitation.* PhD thesis, RWTH Aachen, Aachen, 2001.

[80] S. Jerzembeck, N. Peters, P. Pepiot-Desjardins, and H. Pitsch. Laminar burning velocities at high pressure for primary reference fuels and gasoline: Experimental and numerical investigation. *Combustion and Flame*, 156(2):292–301, 2009.

[81] G. T. Kalghatgi, K. Nakata, and K. Mogi. Octane Appetite Studies in Direct Injection Spark Ignition (DISI) Engines. *SAE Technical Paper*, 2005-01-0244, 2005.

[82] A. Kéromnès, W. K. Metcalfe, K. A. Heufer, N. Donohoe, A. K. Das, C.-J. Sung, J. Herzler, C. Naumann, P. Griebel, O. Mathieu, M. C. Krejci, E. L. Petersen, W. J. Pitz, and H. J. Curran. An experimental and detailed chemical kinetic modeling study of hydrogen and syngas mixture oxidation at elevated pressures. *Combustion and Flame*, 160(6):995–1011, 2013.

[83] M. Klell, H. Eichlseder, and M. Sartory. Mixtures of hydrogen and methane in the internal combustion engine – Synergies, potential and regulations. *International Journal of Hydrogen Energy*, 37(15):11531–11540, 2012.

[84] A. N. Kolmogorov. Dissipation of Energy in Locally Isotropic Turbulence. *Akademiia Nauk SSSR Doklady*, 32:16, 1941.

[85] A. N. Kolmogorov. Dissipation of energy in the locally isotropic turbulence. *Proceedings of the Royal Society of London. Series A - Mathematical and Physical Sciences*, 434(1890):15–17, 1991.

[86] A. A. Konnov. Remaining uncertainties in the kinetic mechanism of hydrogen combustion. *Combustion and Flame*, 152(4):507–528, 2008.

[87] A. A. Konnov, editor. *The Temperature and Pressure Dependences of the Laminar Burning Velocity: Experiments and Modelling*, March 30 - April 2, 2015.

[88] L. D. Landau. On the theory of slow combustion. *Acta Physicochim (USSR)*, 19:77–85, 1944.

[89] C. Law and C. Sung. Structure, aerodynamics, and geometry of pre-mixed flamelets. *Progress in Energy and Combustion Science*, 26(4-6):459–505, 2000.

[90] C. K. Law and F. N. Egolfopoulos. A unified chain-thermal theory of fundamental flammability limits. *Symposium (International) on Combustion*, 24(1):137–144, 1992.

[91] C. K. Law, A. Makino, and T. F. Lu. On the off-stoichiometric peaking of adiabatic flame temperature. *Combustion and Flame*, 145(4):808–819, 2006.

[92] A. N. Lipatnikov and J. Chomiak. Turbulent flame speed and thickness: phenomenology, evaluation and application in multi-dimensional simulations. *Progress in Energy and Combustion Science*, 28(1):1–74, 2002.

[93] C.-C. Liu, S. S. Shy, M.-W. Peng, C.-W. Chiu, and Y.-C. Dong. High-pressure burning velocities measurements for centrally-ignited premixed methane/air flames interacting with intense near-isotropic turbulence at constant Reynolds numbers. *Combustion and Flame*, 159(8):2608–2619, 2012.

[94] J. C. Livengood and P. C. Wu. Correlation of autoignition phenomena in internal combustion engines and rapid compression machines. *Symposium (International) on Combustion*, 5(1):347–356, 1955.

[95] S. Malcher, M. Bargende, M. Grill, U. Baretzky, H. Diel, S. Wohlge-muth, and G. Röttger. Investigation of Flame Propagation Description in Quasi-Dimensional Spark Ignition Engine Modeling. *SAE Technical Paper*, 2018-01-1655, 2018.

[96] V. Manente. *Gasoline partially premixed combustion: An advanced international combustion engine concept aimed to high efficiency, low emissions and low acoustic noise in the whole load range.* PhD Thesis, Lund University, Lund, 2010.

[97] N. M. Marinov. A detailed chemical kinetic model for high temperature ethanol oxidation. *International Journal of Chemical Kinetics*, 31(3):183–220, 1999.

[98] G. H. Markstein. *Nonsteady Flame Propagation*, volume 75 of *AGAR-Dograph*. Elsevier Science, Burlington, 2014.

[99] S. Martinez, A. Irimescu, S. Merola, P. Lacava, and P. Curto-Riso. Flame Front Propagation in an Optical GDI Engine under Stoichiometric and Lean Burn Conditions. *Energies*, 10(9):1337, 2017.

[100] S. Martinez, P. Lacava, P. Curto, A. Irimescu, and S. Merola. Effect of Hydrogen Enrichment on Flame Morphology and Combustion Evolution in a SI Engine Under Lean Burn Conditions. *SAE Technical Paper*, 2018-01-1144, 2018.

[101] M. Matalon. The Darrieus-Landau instability of premixed flames. *Fluid Dynamics Research*, 2018.

[102] Mechanical and Aerospace Engineering (Combustion Research), University of California at San Diego. Chemical-Kinetic Mechanisms for Combustion Applications. http://combustion.ucsd.edu, 13.11.2019.

[103] W. K. Metcalfe, S. M. Burke, S. S. Ahmed, and H. J. Curran. A Hierarchical and Comparative Kinetic Modeling Study of C $1 - $ C 2 Hydrocarbon and Oxygenated Fuels. *International Journal of Chemical Kinetics*, 45(10):638–675, 2013.

[104] R. J. Middleton, J. B. Martz, G. A. Lavoie, A. Babajimopoulos, and D. N. Assanis. A computational study and correlation of premixed isooctane air laminar reaction fronts diluted with EGR. *Combustion and Flame*, 159(10):3146–3157, 2012.

[105] G. Mittal, S. M. Burke, V. A. Davies, B. Parajuli, W. K. Metcalfe, and H. J. Curran. Autoignition of ethanol in a rapid compression machine. *Combustion and Flame*, 161(5):1164–1171, 2014.

[106] M.T. Nguyen, L.J. Jiang, and Shenqyang (Steven) Shy. Fuel Similarity and Turbulent Burning Velocities of Stoichiometric Iso-octane, Lean Hydrogen, and Lean Propane at High Pressure. In *26th ICDERS, USA, July 30 - August 4, 2017*.

[107] U. C. Müller, M. Bollig, and N. Peters. Approximations for burning velocities and markstein numbers for lean hydrocarbon and methanol flames. *Combustion and Flame*, 108(3):349–356, 1997.

[108] G. Müller-Syring and M. Henel. Wasserstofftoleranz der Erdgas-infrastruktur inklusive aller assoziierten Anlagen: Abschlussbericht DVGW Forschung.

[109] S. Muppala, N. Aluri, F. Dinkelacker, and A. Leipertz. Development of an algebraic reaction rate closure for the numerical calculation of turbulent premixed methane, ethylene, and propane/air flames for pressures up to 1.0 MPa. *Combustion and Flame*, 140(4):257–266, 2005.

[110] A. Nordelöf, M. Messagie, A.-M. Tillman, M. Ljunggren Söderman, and J. van Mierlo. Environmental impacts of hybrid, plug-in hybrid, and battery electric vehicles—what can we learn from life cycle assessment? *The International Journal of Life Cycle Assessment*, 19(11):1866–1890, 2014.

[111] R. Nußbaumer. *Reaction kinetics calculation of laminar flame speeds of gasoline fuels with different octane number and ethanol content*. Master's thesis, University of Stuttgart, Stuttgart, 2017.

[112] E. C. Okafor, Y. Nagano, and T. Kitagawa. Effects of Hydrogen Concentration on Stoichiometric $H_2/CH_4/Air$ Premixed Turbulent Flames. *SAE Technical Paper*, 2013-01-2563, 2013.

[113] R. H. Pahl and M. J. McNally. Fuel Blending and Analysis for the Auto/Oil Air Quality Improvement Research Program. *SAE Technical Paper*, 902098, 1990.

[114] B. Pang, M.-Z. Xie, M. Jia, and Y.-D. Liu. Development of a Phenomenological Soot Model Coupled with a Skeletal PAH Mechanism for Practical Engine Simulation. *Energy & Fuels*, 27(3):1699–1711, 2013.

[115] F. Perini, K. Zha, S. Busch, E. Kurtz, R. C. Peterson, A. Warey, and R. D. Reitz. Piston geometry effects in a light-duty, swirl-supported diesel engine: Flow structure characterization. *International Journal of Engine Research*, 19(10):1079–1098, 2018.

[116] N. Peters. Fifteen Lectures on Laminar and Turbulent Combustion: Summer School, 1992. https://www.itv.rwth-aachen.de/fileadmin/Downloads/Summerschools/SummerSchool.pdf, 19.05.2020.

[117] N. Peters. The turbulent burning velocity for large-scale and small-scale turbulence. *Journal of Fluid Mechanics*, 384:107–132, 1999.

[118] N. Peters. Kinetic Foundation of Thermal Flame Theory. In E. A. Thornton, editor, *Advances in Combustion Science*, volume 12, pages 73–91. American Institute of Aeronautics and Astronautics, Reston, 2000.

[119] N. Peters. *Turbulent Combustion*. Cambridge University Press, Cambridge, 2000.

[120] F. Pischinger. Sonderforschungsbereich 224 Motorische Verbrennung, 2001. http://www.sfb224.rwth-aachen.de/bericht.htm, 13.11.2019.

[121] T. Poinsot and D. Veynante. *Theoretical and numerical combustion*. Edwards, Philadelphia, 2. ed. edition, 2005.

[122] S. B. Pope. *Turbulent Flows*. Cambridge University Press, 2000.

[123] Prof. Dr.-Ing. Ulrich Brill, editor. *Klopfregelung für Ottomotoren II: Selbstzündung im Klopfgrenzbereich von Serienmotoren*, volume 74. expert-Verlag, Essen, 2006.

[124] R. T. E. Hermanns, R. J. M. Bastiaans, L. P. H. De Goey. Asymptotic Analysis of Methane-Hydrogen-Air Mixtures. In *Proceedings of the European Combustion Meeting : Louvain-la-Neuve, Belgium, April 3 - 6, 2005.*

[125] E. Ranzi, A. Frassoldati, S. Granata, and T. Faravelli. Wide-Range Kinetic Modeling Study of the Pyrolysis, Partial Oxidation, and Combustion of Heavy n -Alkanes. *Industrial & Engineering Chemistry Research*, 44(14):5170–5183, 2005.

[126] A. Ratzke, T. Schöffler, K. Kuppa, and F. Dinkelacker. Validation of turbulent flame speed models for methane–air-mixtures at high pressure gas engine conditions. *Combustion and Flame*, 162(7):2778–2787, 2015.

[127] S. Ravi. *Measurement of Turbulent Flame Speeds of Hydrogen and Natural Gas Blends (C1-C5 Alkanes) using a Newly Developed Fan-Stirred Vessel: Ph.D. thesis.* Texas A & M University, 2014.

[128] M. G. Reid and R. Douglas. Quasi-Dimensional Modelling of Combustion in a Two-Stroke Cycle Spark Ignition Engine. *SAE Technical Paper*, 941680, 1994.

[129] S. Richard, S. Bougrine, G. Font, F.-A. Lafossas, and F. Le Berr. On the Reduction of a 3D CFD Combustion Model to Build a Physical 0D Model for Simulating Heat Release, Knock and Pollutants in SI Engines. *Oil & Gas Science and Technology - Revue de l'IFP*, 64(3):223–242, 2009.

[130] J.-Y. Robin and V. Demoury. The LNG Industry in 2012. *International Group of Liquefied Natural Gas Importers*, 2012.

[131] M. Schmid. *Investigation of the influence of ctual combustion chamber geometries on the calculation of flame surfaces and resulting burn rate changes in engine simulation.* Student research project, University of Stuttgart, Stuttgart, 2018.

[132] L. Sileghem, J. Vancoillie, J. Demuynck, J. Galle, and S. Verhelst. Alternative Fuels for Spark-Ignition Engines: Mixing Rules for the Laminar Burning Velocity of Gasoline–Alcohol Blends. *Energy & Fuels*, 26(8):4721–4727, 2012.

[133] R. J. Tabaczynski, C. R. Ferguson, and K. Radhakrishnan. A Turbulent Entrainment Model for Spark-Ignition Engine Combustion. *SAE Technical Paper*, 770647, 1977.

[134] G. I. Taylor and A. E. Green. Mechanism of the production of small eddies from large ones. *Proceedings of the Royal Society of London. Series A - Mathematical and Physical Sciences*, 158(895):499–521, 1937.

[135] H. Tennekes and J. L. Lumley. *A first course in turbulence*. MIT Press, Cambridge, Massachusetts and London, England, 1972.

[136] S. R. Turns. *An introduction to combustion: Concepts and applications*. McGraw-Hill series in mechanical engineering. McGraw-Hill, New York, 3rd ed. edition, 2012.

[137] L. Urban, M. Grill, S. Hann, and M. Bargende. Simulation of Autoignition, Knock and Combustion for Methane-Based Fuels. *SAE Technical Paper*, 2017-01-2186, 2017.

[138] J. P. van Lipzig. *Flame speed investigation of ethanol, n-heptane and iso-octane using the heat flux method*. Master's thesis, Lund Institute of Technology, Lund, 2010.

[139] J. Vancoillie, J. Demuynck, J. Galle, S. Verhelst, and J. van Oijen. A laminar burning velocity and flame thickness correlation for ethanol-air mixtures valid at spark-ignition engine conditions. *Fuel*, 102:460–469, 2012.

[140] S. Verhelst and C. Sheppard. Multi-zone thermodynamic modelling of spark-ignition engine combustion – An overview. *Energy Conversion and Management*, 50(5):1326–1335, 2009.

[141] I. I. Vibe. *Brennverlauf und Kreisprozeß von Verbrennungsmotoren*. Verl. Technik, Berlin, 1970.

[142] D. Vuilleumier and M. Sjöberg. Significance of RON, MON, and LTHR for Knock Limits of Compositionally Dissimilar Gasoline Fuels in a DISI Engine. *SAE International Journal of Engines*, 10(3):938–950, 2017.

[143] T. Wabel, A. Skiba, J. Temme, and J. Driscoll. Measurements to determine the regimes of premixed flames in extreme turbulence. *Proceedings of the Combustion Institute*, 36(2):1809–1816, 2017.

[144] M. Weiß, N. Zarzalis, and R. Suntz. Experimental study of Markstein number effects on laminar flamelet velocity in turbulent premixed flames. *Combustion and Flame*, 154(4):671–691, 2008.

[145] G. A. Weisser. *Modelling of combustion and nitric oxide formation for medium-speed DI diesel engines: a comparative evaluation of zero- and three-dimensional approaches.* PhD thesis, ETH Zurich, 2001.

[146] M. Wenig. *Simulation der ottomotorischen Zyklenschwankungen: Ph.D. thesis.* Department Konstruktions-, Produktions- und Fahrzeugtechnik, University of Stuttgart, 2013.

[147] M. Wenig, M. Grill, and M. Bargende. A New Approach for Modeling Cycle-to-Cycle Variations within the Framework of a Real Working-Process Simulation. *SAE International Journal of Engines*, 6(2):1099–1115, 2013.

[148] C. Wouters, B. Lehrheuer, B. Heuser, and S. Pischinger. Gasoline Blends with Methanol, Ethanol and Butanol. *MTZ*, 81(3):16–21, 2020.

[149] Q. Yang, M. Grill, and M. Bargende. A Quasi-Dimensional Charge Motion and Turbulence Model for Diesel Engines with a Fully Variable Valve Train. *SAE Technical Paper*, 2018-01-0165, 2018.

[150] S. Yang, A. Saha, Z. Liu, and C. K. Law. Role of Darrieus–Landau instability in propagation of expanding turbulent flames. *Journal of Fluid Mechanics*, 850:784–802, 2018.

[151] Y. B. Zeldovich and D. A. Frank-Kamenetskii. The Theory of Thermal Propagation of Flames. *Russian Journal of Physical Chemistry A*, 12:100–105, 1938.

[152] M. Zhang, J. Wang, Y. Xie, W. Jin, Z. Wei, Z. Huang, and H. Kobayashi. Flame front structure and burning velocity of turbulent premixed CH4/H2/air flames. *International Journal of Hydrogen Energy*, 38(26):11421–11428, 2013.

[153] V. L. Zimont. Gas premixed combustion at high turbulence. Turbulent flame closure combustion model. *Experimental Thermal and Fluid Science*, 21(1-3):179–186, 2000.

Appendix

A.1 Overview: Laminar Flame Speed of All Investigated Fuels

Figure A1.1 compares the laminar flame speed s_L of all fuels investigated. s_L of methyl formate and DMC+ were calculated using the reaction mechanism from [23]. The reaction mechanism published by *Cai et al.* [22] was used for all other fuels. The qualitative s_L-difference between gasoline and ethanol is supported by measurements published in [34] and [6]. Measurements performed by *Farrell et al.* [45] prove the s_L-difference between methane and ethane and the similarity of ethane and ethanol. Measurements published in [7] show a higher s_L for iso-octane, compared to methane.

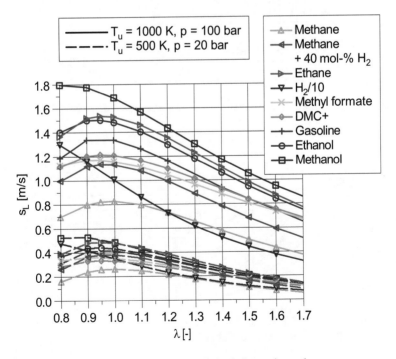

Figure A1.1: Laminar flame speed of all fuels investigated

A.2 Additional Investigations of the Laminar Flame Speed

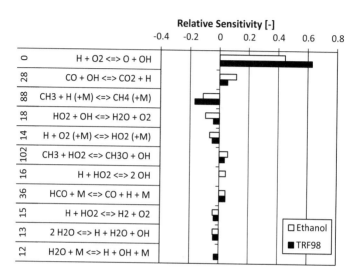

Figure A2.1: Sensitivity analysis of ethanol and gasoline (TRF98) [71]

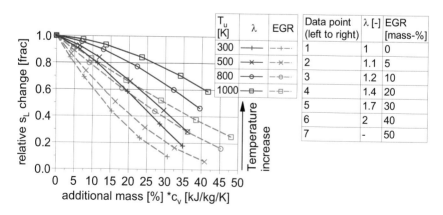

Figure A2.2: Relative laminar flame speed change with λ and EGR rate, x-axis scaling includes heat capacity influence, p = 1 bar [71]

A.3 Measurement Data Analysis at Different Engine Speeds

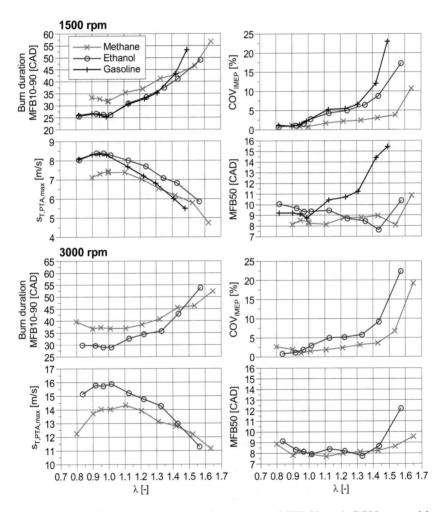

Figure A3.1: Change in burn duration, $s_{T,PTA}$, MFB50 and COV$_{IMEP}$ with changing λ and fuel, 1500 rpm and 3000 rpm, engine A

A.4 s_L-Models of *Gülder* and *Heywood* vs. Reaction Kinetics Calculations

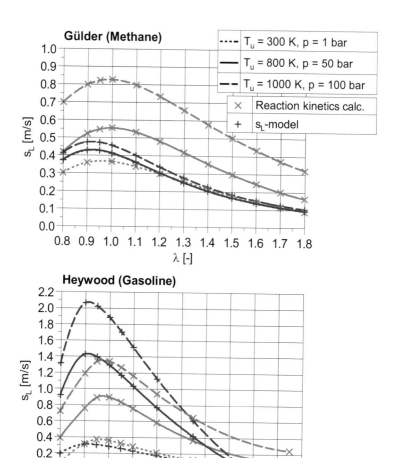

Figure A4.1: Comparison of the *Gülder* and *Heywood* models with laminar flame speeds from reaction kinetics. Bottom graph from [71] (edited)

A.5 Calibration Parameters of the Laminar Flame Speed Model

Table A5.1: Calibration parameters of the laminar flame speed model: gaseous fuels

	Methane	Ethane	Methane + 40 % H_2	Hydrogen
L_{min}	17.167	16.6	18.529	34.306
Z_{st}^*	0.055	0.057	0.051	0.028
E_i [K]	52686.3	26215.7	36938.6	56500.0
B_i [bar]	1.229E+18	6.039E+09	6.446E+11	1.230E+18
m	1.5	1.5	1.5	1.5
r	0.985	0.985	0.985	0.985
n	2.439	2.439	2.439	2.439
F [m/s]	0.00017136	1.46149	0.10044564	0.000088035419
G [K]	-15155.10	-2046.95	-7205.81	-19332.52
n_{EGR}	1.991	1.390	1.807	1.622
n_a	1.020	0.945	0.945	2.000
c	0.639	0.748	0.594	0.739
b_{yH2O}	0.737	0.698	0.733	0.367
c_{yH2O} [K]	2833.7	2497.3	2497.3	5268.6
d_{yH2O}	10.189	10.180	10.180	10.189

Table A5.2: Calibration parameters of the laminar flame speed model: liquid fuels

	Gasoline (TRF)	Methanol	Ethanol	Methyl formate	DMC+
L_{min}	14.7	6.75	9	4.5737	4.8
Z_{st}^{*}	0.064	0.129	0.100	0.179	0.172
E_i [K]	53361.3	52457.3	56760.4	37518.5	46854.0
B_i [bar]	1.229E+18	1.229E+18	1.229E+18	3.585E+12	2.141E+15
m	1.5	1.5	1.5	1.5	1.5
r	0.985	0.985	0.985	0.973	0.973
n	2.439	2.439	2.439	2.442	2.442
F [m/s]	0.0027796	0.0394036	0.0042784	0.595045	0.2000456
G [K]	-10607.25	-6819.61	-11066.26	-3535.99	-5178.14
n_{EGR}	1.499	1.184	0.890	1.098	0.771
n_a	0.945	1.000	0.902	0.902	0.902
c	0.748	0.721	0.820	0.676	0.820
b_{yH2O}	0.681	1.181	1.002	1.542	1.562
c_{yH2O} [K]	2756.2	1496.6	1908.0	1227.2	1175.3
d_{yH2O}	10.189	10.189	10.189	1E+18	1E+09

Table A5.3: Calibration parameters of the laminar flame speed model: splines for methane and gasoline

λ	Methane				Gasoline (TRF)			
	S1 [-]	S2 [-]	S3 [K]	S4 [-]	S1 [-]	S2 [-]	S3 [K]	S4 [-]
0.6	0.0000	1.111420	1838.7	0.6171	0.0000	1.017912	1412.1	0.7688
0.7	0.0000	1.060078	1790.1	0.2712	0.0000	1.017446	1522.1	0.8084
0.8	0.0000	1.058724	1763.5	0.3995	0.0000	0.996815	1707.8	0.6482
0.9	0.0000	1.056041	1789.4	0.4677	0.0000	0.995619	1889.3	0.5837
0.95	0.0000	1.053646	1810.2	0.4526	0.0000	0.996434	1944.4	0.5438
1	0.0000	1.061577	1812.3	0.5803	0.0000	0.995548	1958.0	0.5090
1.1	0.0000	1.052061	1799.7	0.4414	0.0000	0.996459	1931.5	0.4932
1.2	0.0000	1.052923	1763.0	0.4388	0.0000	1.000830	1847.1	0.5800
1.3	0.0180	1.035266	1639.1	0.4661	0.0176	0.983174	1718.5	0.5391
1.4	0.0253	1.031419	1536.8	0.5380	0.0288	0.972679	1611.7	0.5532
1.5	0.0410	1.016737	1447.5	0.5515	0.0409	0.960939	1521.3	0.5446
1.6	0.0533	1.003558	1385.2	0.5399	0.0542	0.948393	1433.8	0.5527
1.7	0.0655	0.995630	1308.2	0.5951	0.0689	0.931626	1366.8	0.5353
1.8	0.0745	1.004773	1246.5	0.7140	0.0818	0.920171	1300.1	0.5423
1.9	0.0975	0.994729	1190.8	0.7647	0.0946	0.907679	1240.6	0.5499
2	0.1108	0.999415	1140.2	0.8599	0.1074	0.891913	1180.4	0.5563
2.1	0.1240	1.012783	1094.1	0.9618	0.1203	0.875526	1127.4	0.5631
2.2	0.1372	1.021877	1051.9	1.0667	0.1331	0.859768	1077.0	0.5693
2.3	0.1503	1.032493	1013.0	1.1780	0.1460	0.846646	1038.1	0.5773
2.4	0.1633	1.044033	977.2	1.2952	0.1588	0.831071	997.1	0.5841
2.5	0.1762	1.054775	944.0	1.4086	0.1716	0.820481	971.1	0.5943
2.6	0.1891	1.067123	913.2	1.5345	0.1845	0.808074	941.7	0.6032
2.7	0.2024	1.080289	884.5	1.6721	0.1973	0.797971	919.3	0.6141
2.8	0.2154	1.093274	857.7	1.8143	0.2101	0.778220	880.7	0.6188
2.9	0.2285	1.101667	832.6	1.9312	0.2230	0.764619	855.3	0.6278
3	0.2416	1.106909	809.1	2.0286	0.2358	0.743352	815.5	0.6356
4	0.3724	1.110354	634.3	2.7856	0.3642	0.637878	674.5	0.7171
5	0.5032	1.103312	525.2	3.7850	0.4925	0.510704	508.1	0.7986

Table A5.4: Calibration parameters of the laminar flame speed model: splines for ethanol and methanol

λ	Ethanol				Methanol			
	S1 [-]	S2 [-]	S3 [K]	S4 [-]	S1 [-]	S2 [-]	S3 [K]	S4 [-]
0.6	0.0000	0.980012	1550.6	0.5742	0.0000	0.945807	1472.7	0.5660
0.7	0.0000	0.988667	1698.8	0.5667	0.0000	0.954612	1626.9	0.5603
0.8	0.0000	0.985877	1826.4	0.5577	0.0000	0.980074	1912.6	0.5624
0.9	0.0000	0.981826	1948.2	0.4375	0.0000	0.980642	2000.3	0.4401
0.95	0.0000	0.980616	1968.1	0.4040	0.0000	0.980863	1995.0	0.4063
1	0.0000	0.979267	1969.9	0.3757	0.0000	0.981086	1983.6	0.3796
1.1	0.0000	0.977698	1921.7	0.3751	0.0000	0.981319	1918.2	0.3775
1.2	0.0000	0.974361	1864.1	0.3527	0.0000	0.981322	1861.3	0.3559
1.3	0.0181	0.953484	1726.1	0.3617	0.0181	0.964490	1739.8	0.3667
1.4	0.0296	0.952086	1643.4	0.4516	0.0296	0.957634	1613.9	0.4539
1.5	0.0420	0.939564	1556.3	0.4628	0.0419	0.947557	1534.3	0.4669
1.6	0.0556	0.923839	1482.7	0.4528	0.0556	0.934281	1466.8	0.4589
1.7	0.0708	0.905626	1414.1	0.4405	0.0687	0.926067	1372.8	0.5328
1.8	0.0840	0.889312	1358.1	0.4295	0.0815	0.914328	1314.0	0.5406
1.9	0.0971	0.871534	1307.6	0.4184	0.0943	0.901560	1258.0	0.5485
2	0.1068	0.867454	1295.0	0.4159	0.1071	0.889128	1208.8	0.5563
2.1	0.1070	0.884767	1221.0	0.5617	0.1200	0.876699	1164.6	0.5642
2.2	0.1198	0.870874	1173.4	0.5702	0.1328	0.863591	1122.3	0.5721
2.3	0.1325	0.856707	1129.8	0.5789	0.1457	0.850118	1082.3	0.5800
2.4	0.1453	0.844433	1094.4	0.5889	0.1585	0.836240	1044.4	0.5879
2.5	0.1581	0.832617	1062.5	0.5993	0.1714	0.822121	1008.2	0.5958
2.6	0.1709	0.821430	1033.9	0.6103	0.1843	0.807935	974.0	0.6039
2.7	0.1837	0.809038	1004.3	0.6209	0.1972	0.794195	943.3	0.6119
2.8	0.1966	0.797394	978.1	0.6320	0.2101	0.780764	914.7	0.6203
2.9	0.2094	0.785569	952.3	0.6434	0.2230	0.767569	888.5	0.6287
3	0.2223	0.776004	932.0	0.6564	0.2361	0.754201	863.5	0.6367
4	0.3474	0.626265	676.1	0.7360	0.3637	0.631314	669.0	0.7211
5	0.3445	0.600718	643.2	0.7051	0.5023	0.534706	574.0	0.7838

Table A5.5: Calibration parameters of the laminar flame speed model: splines for methyl formate and DMC+

	Methyl formate				DMC+			
λ	S1 [-]	S2 [-]	S3 [K]	S4 [-]	S1 [-]	S2 [-]	S3 [K]	S4 [-]
0.6	0.0000	1.030361	1734.4	0.5931	0.0000	1.049590	1722.0	0.5920
0.7	0.0000	1.017155	1862.3	0.5808	0.0000	1.041270	1847.1	0.5796
0.8	0.0000	1.008536	1946.3	0.5645	0.0000	1.026395	1935.2	0.5637
0.9	0.0000	0.998629	2025.9	0.4398	0.0000	1.012417	2016.6	0.4395
0.95	0.0000	0.994731	2030.1	0.4052	0.0000	1.007314	2021.3	0.4049
1	0.0000	0.989994	2017.7	0.3725	0.0000	1.001620	2009.4	0.3723
1.1	0.0000	0.989973	1978.8	0.3763	0.0000	0.999523	1971.9	0.3761
1.2	0.0000	0.983540	1914.5	0.3541	0.0000	0.992054	1908.2	0.3539
1.3	0.0181	0.956612	1774.4	0.3631	0.0181	0.965006	1768.4	0.3629
1.4	0.0296	0.954264	1673.5	0.4531	0.0296	0.961262	1668.5	0.4528
1.5	0.0420	0.940734	1590.6	0.4654	0.0420	0.947388	1585.8	0.4651
1.6	0.0557	0.917842	1500.2	0.4548	0.0558	0.924654	1495.3	0.4545
1.7	0.0709	0.898599	1436.6	0.4433	0.0710	0.905216	1431.6	0.4429
1.8	0.0861	0.880960	1385.8	0.4327	0.0862	0.887509	1380.7	0.4323
1.9	0.1013	0.856067	1319.3	0.4211	0.1014	0.862737	1313.9	0.4207
2	0.1164	0.821331	1234.8	0.4080	0.1164	0.828378	1228.9	0.4075
2.1	0.1315	0.801101	1194.8	0.3988	0.1316	0.808053	1188.7	0.3982
2.2	0.1483	0.776544	1152.4	0.3868	0.1484	0.783477	1146.1	0.3862
2.3	0.1625	0.750109	1105.9	0.3755	0.1627	0.757079	1099.3	0.3749
2.4	0.1768	0.722721	1061.5	0.3642	0.1770	0.729286	1055.0	0.3635
2.5	0.1911	0.695008	1019.9	0.3528	0.1914	0.701215	1013.4	0.3522
2.6	0.2075	0.664808	980.4	0.3387	0.2078	0.670538	974.1	0.3381
2.7	0.2248	0.632760	939.0	0.3273	0.2252	0.637946	932.7	0.3267
2.8	0.2404	0.597509	894.6	0.3126	0.2408	0.602047	888.7	0.3120
2.9	0.2589	0.573477	873.0	0.3029	0.2593	0.577268	866.7	0.3023
3	0.2706	0.523629	800.0	0.2909	0.2711	0.526555	794.6	0.2903
4	0.4118	0.385854	773.0	0.1823	0.4125	0.387160	770.0	0.1821
5	0.5539	0.241597	747.0	0.0739	0.5539	0.243000	747.0	0.0739

Table A5.6: Calibration parameters of the laminar flame speed model: splines for ethane and hydrogen

λ	Ethane				Hydrogen			
	S1 [-]	S2 [-]	S3 [K]	S4 [-]	S1 [-]	S2 [-]	S3 [K]	S4 [-]
0.6	0.0000	1.100049	1807.0	0.6001	0.0000	0.925668	1672.9	0.5867
0.7	0.0000	1.063385	1829.9	0.5783	0.0000	0.945232	1822.3	0.5773
0.8	0.0000	1.025248	1890.5	0.5610	0.0000	0.959804	1973.1	0.5655
0.9	0.0000	1.004117	1991.8	0.4387	0.0000	0.961368	1959.3	0.4393
0.95	0.0000	0.998181	1999.1	0.4043	0.0000	0.959482	1896.5	0.4031
1	0.0000	0.991791	1984.9	0.3717	0.0000	0.968262	1853.5	0.3740
1.1	0.0000	0.991274	1942.3	0.3754	0.0000	0.955160	1742.7	0.3714
1.2	0.0000	0.984919	1872.4	0.3532	0.0000	0.951248	1668.9	0.3481
1.3	0.0181	0.958935	1732.5	0.3620	0.0181	0.935692	1569.8	0.3591
1.4	0.0296	0.959823	1637.8	0.4515	0.0296	0.928933	1465.4	0.4436
1.5	0.0421	0.951907	1569.7	0.4648	0.0420	0.919140	1406.2	0.4550
1.6	0.0559	0.939372	1507.9	0.4563	0.0596	0.913408	1362.3	0.4612
1.7	0.0713	0.924717	1450.9	0.4457	0.0716	0.890929	1313.2	0.4365
1.8	0.0844	0.921164	1431.5	0.4363	0.0859	0.888949	1328.7	0.4310
1.9	0.0978	0.908959	1389.5	0.4262	0.1002	0.871744	1277.1	0.4211
2	0.1114	0.894751	1348.1	0.4147	0.1137	0.878921	1331.5	0.4110
2.1	0.1319	0.825723	1183.1	0.3977	0.1304	0.835619	1185.2	0.4149
2.2	0.1488	0.802636	1142.5	0.3859	0.1466	0.818468	1137.6	0.4287
2.3	0.1632	0.777161	1097.4	0.3747	0.1500	0.815297	1126.8	0.4261
2.4	0.1777	0.749406	1052.3	0.3633	0.1632	0.800799	1091.1	0.4364
2.5	0.1921	0.718877	1005.9	0.3515	0.1764	0.786887	1057.4	0.4467
2.6	0.2086	0.684982	961.1	0.3369	0.1897	0.773039	1026.4	0.4571
2.7	0.2263	0.654330	924.7	0.3259	0.2030	0.759393	997.4	0.4677
2.8	0.2415	0.599957	846.0	0.3116	0.2163	0.745024	968.8	0.4781
2.9	0.2523	0.573417	815.3	0.3006	0.2320	0.741470	967.6	0.4884
3	0.2683	0.553378	801.2	0.2916	0.2453	0.730359	947.8	0.4988
4	0.4313	0.396324	780.0	0.1734	0.3785	0.631365	796.0	0.6028
5	0.5718	0.235566	747.0	0.0514	0.5117	0.552552	695.1	0.7068

Table A5.7: Calibration parameters of the laminar flame speed model: splines for methane + 40 mol−% H_2

	Methane + 40 % H_2			
λ	S1 [-]	S2 [-]	S3 [K]	S4 [-]
0.6	0.0000	1.033046	1734.1	0.5932
0.7	0.0000	1.040977	1844.7	0.5795
0.8	0.0000	1.027111	1893.1	0.5612
0.9	0.0000	1.009312	1988.4	0.4386
0.95	0.0000	1.003336	1997.0	0.4043
1	0.0000	0.996405	1983.2	0.3717
1.1	0.0000	0.993188	1942.3	0.3754
1.2	0.0000	0.984124	1873.4	0.3532
1.3	0.0028	0.986742	1812.6	0.4088
1.4	0.0056	0.984457	1745.8	0.4304
1.5	0.0083	0.984300	1685.0	0.4585
1.6	0.0111	0.986883	1632.7	0.4925
1.7	0.0139	0.992602	1585.4	0.5376
1.8	0.0167	1.002890	1546.1	0.5958
1.9	0.0194	1.016683	1510.9	0.6648
2	0.0222	1.035619	1479.6	0.7524
2.1	0.0250	1.066083	1447.9	0.8442
2.2	0.0278	1.087650	1418.1	0.9351
2.3	0.0306	1.110110	1390.2	1.0283
2.4	0.0333	1.133345	1364.0	1.1240
2.5	0.0361	1.157007	1339.3	1.2211
2.6	0.0389	1.181362	1316.1	1.3203
2.7	0.0417	1.205997	1294.0	1.4212
2.8	0.0444	1.231147	1273.2	1.5245
2.9	0.0472	1.257018	1253.4	1.6315
3	0.0500	1.284745	1234.5	1.7457
4	0.0778	1.408194	1085.5	2.2913
5	0.1056	1.768607	982.5	3.4637

A.6 Laminar Flame Speed Model Quality

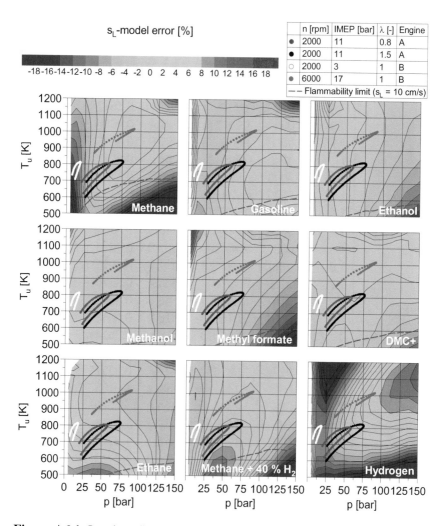

Figure A6.1: Laminar flame speed model error at $\lambda = 1.6$ [72] (updated and expanded)

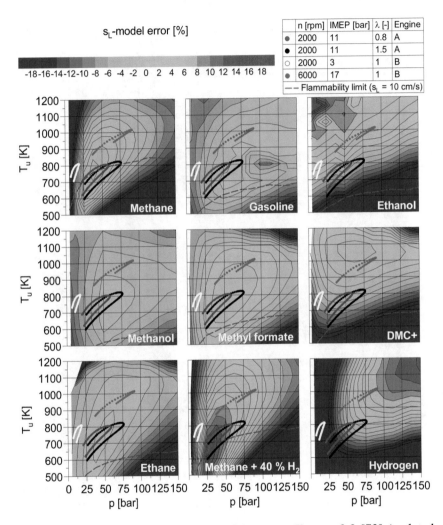

Figure A6.2: Laminar flame speed model error at $Y_{EGR} = 0.3$ [72] (updated and expanded)

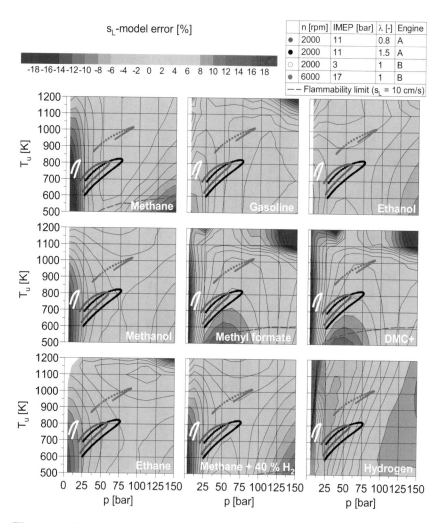

Figure A6.3: Laminar flame speed model error at $Y_{H2O} = 1$ [72] (updated and expanded)

A.7 Comparison of Flame Surface Area Calculation Methods

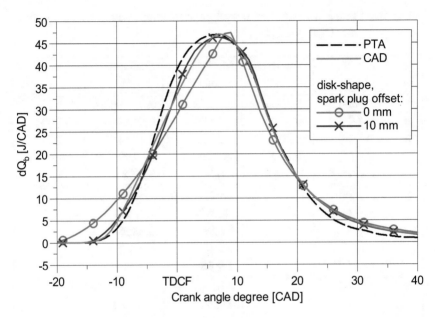

Figure A7.1: Detailed flame surface versus simplified approach [72] (edited)

A.8 Engine Knock: Additional Model Parameters and Figures

Table A8.1: Ignition delay time model parameters of methanol, methyl formate, DMC+ and H_2

τ_i [ms]	Param.	Methanol	Methyl formate	DMC+	H_2
τ or τ_1	A [ms]	4.037E-12	9.8657E-13	2.7771E-16	1.061E-12
	α [-]	-0.8315	0.9043	0.9043	-0.8000
	β [-]	0.4395	3.2222	2.1354	0.5522
	γ [-]	-0.4897	-3.7709	0.0299	-0.6901
	E_A/R [K]	19418	24881	33818	21174
τ_2	A [ms]	-	1.9271E-06	7.6848E-06	-
	α [-]	-	-1.1859	-1.1859	-
	β [-]	-	0.9585	0.3856	-
	γ [-]	-	-1.1478	-0.4216	-
	E_A/R [K]	-	13895	12851	-
τ_3	A [ms]	-	4.6150E-08	1.2851E-04	-
	α [-]	-	-0.8942	-0.8942	-
	β [-]	-	0.5591	0.7087	-
	γ [-]	-	-0.9315	-1.1544	-
	E_A/R [K]	-	16572	17299	-

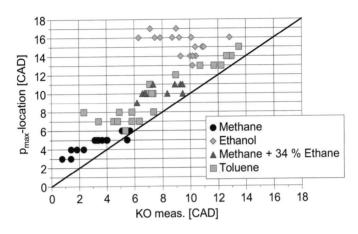

Figure A8.2: Comparison: measured knock onset versus peak pressure location, engine A

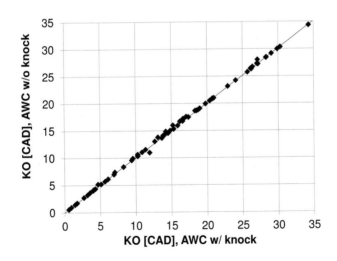

Figure A8.3: Knock onset comparison: averaged working cycle calculated with and without knocking single working cycles, engine A

Printed in the United States
By Bookmasters